Sxi – Springer per l'Innovazione

Sxi – Springer for Innovation

Peregrina Quintela • Ana Belén Fernández •
Adela Martínez • Guadalupe Parente •
María Teresa Sánchez

TransMath

Innovative Solutions from Mathematical Technology

 Springer

Peregrina Quintela
Department of Applied Mathematics
University of Santiago de Compostela
Santiago de Compostela, Spain

Ana Belén Fernández
Department of Applied Mathematics
University of Santiago de Compostela
Santiago de Compostela, Spain

Adela Martínez
Applications & Projects
CESGA Foundation
Santiago de Compostela, Spain

Guadalupe Parente
Applications & Projects
CESGA Foundation
Santiago de Compostela, Spain

María Teresa Sánchez
Applications & Projects
CESGA Foundation
Santiago de Compostela, Spain

Sxi – Springer per l'Innovazione / Sxi – Springer for Innovation
ISSN: 2239-2688   ISSN: 2239-2696 (electronic)
ISBN 978-88-470-2405-2   ISBN 978-88-470-2406-9 (eBook)
DOI 10.1007/978-88-470-2406-9
Springer Milan Heidelberg New York Dordrecht London

Library of Congress Control Number: 2012938115

9 8 7 6 5 4 3 2 1

Cover design: Beatrice Ꞵ, Milano
Cover image: Alba Souto

Typesetting: PTP-Berlin, Protago TeX-Production GmbH, Germany (www.ptp-berlin.eu)

Springer-Verlag Italia S.r.l., Via Decembrio 28, I-20137 Milano
Springer-Verlag is part of Springer Science+Business Media (www.springer.com)

Preface

The book *TransMath: Innovative Solutions from Mathematical Technology* has been conceived as a powerful tool to spread and champion the transfer of mathematical knowledge and techniques to the industrial sector and society. This publication presents two technological maps, developed in the Consolider i-MATH CSD2006-00032 Project (Ministry of Science and Innovation – Government of Spain). Firstly, the i-MATH Map of Demand for Mathematical Technology (TransMATH Demand Map), with data obtained from a survey carried out with around 5,200 Spanish firms, details the level of knowledge, use and demand for mathematical technology in industry[1] in Spain. Secondly, the i-MATH Map of Supply of Mathematical Technology (TransMATH Supply Map) shows the experience acquired in technology transfer to business and industrial sectors by a broad representation of those Spanish mathematical research groups with greatest capacity and background in collaborating with industry.

The book is mainly addressed to those companies with innovation needs that could be met using mathematical technology. A complete list of successful industrial projects, developed by Spanish research groups, is therefore included to help readers determine the level of implementation and demand for mathematical technology in other companies and sectors of economic activity. Furthermore, it illustrates the benefits of using mathematical techniques to enhance innovation in industry.

In regards to professional readers, all information collected in this publication is classified by economic activity. Eleven sectors are considered in particular, as the majority of supply and demand for mathematical technology corresponds to companies belonging to these sectors: *Biomedicine & Health, Construction, Economics & Finance, Energy & Environment, Food, Information & Communication Technology (ICT), Logistics & Transport, Management & Tourism, Metal & Machinery, Public Administration,* and *Technical Services*.

---

[1] In this book the term "industry" is used in a broad sense of the word, to denote all kinds of business and commercial firms with economic activity and non-profit R&D organisations with activities outside the realm of education and academic research (including financial institutions, public administration and hospitals).

For readers interested in state of the art Spanish mathematical technology transferred to business and industrial sectors, two dedicated websites http://www.math-in.net, and http://www.i-math.org/ provide further information.

Santiago de Compostela, January 2012

Peregrina Quintela
Ana Belén Fernández
Adela Martínez
Guadalupe Parente
María Teresa Sánchez

Acknowledgments

Funding for the technological maps compiled in this book has come from the Spanish Ministry of Science and Education through the Consolider i-MATH CSD2006-00032 Project and from the Complementary Action MTM2007-30179-E, from the Xunta de Galicia (Regional Government of Galicia) via the agreement dated 16/10/2007, from the Red Mathematica Consulting & Computing de Galicia, from the Centro de Supercomputación de Galicia (CESGA) and from the Universidade de Santiago de Compostela (USC).

Furthermore, the foundations of this book are built upon earlier editions of the publications *Investigadores en matemáticas para dar soluciones innovadoras (TransMATH Oferta)* and *Mapa i-MATH de demanda empresarial de tecnología matemática (TransMATH Demanda)*. The authors wish to thank all those who have contributed to these publications for their invaluable help in producing this book: Aureli Alabert, Mª Teresa Alonso, Alfredo Bermúdez de Castro, Javier Bullón, Laureano Escudero, Mª Jesús Frieiro, Jesús Gil, Mª José Ginzo, Andrés Gómez, Wenceslao González, Daniel Hernández, Mónica López, Félix Martínez, Jeff Palmer, Sophie Elizabeth Paton, Julio Rubio, Mª Teresa Seoane, Arturo Soto, Alba Souto and Giuseppe Viglialoro.

Contents

Introduction

1

The research project Ingenio Mathematica (i-MATH), an initiative financed by the Spanish Ministry of Education and Science, promoted a wide range of activities for the period 2006–2012 in order to improve the role of mathematical research in the Spanish sphere of science, technology and innovation, and increase the transferal of knowledge from Spanish mathematicians to business sectors. Specifically, two initiatives were developed in order to analyse the current state of mathematical technology transfer in Spain: the TransMATH Demand Map, which studies the knowledge, use and requirements for mathematical tools by Spanish companies, and the TransMATH Supply Map, which compiles both Spanish research groups' experience in technology transfer and the key mathematical services they currently offer.

1.1
Background

During the period 2006–2012 the mathematical community in Spain has contributed to a new global initiative aimed at promoting knowledge transfer from universities and public research bodies to industry. Thus, the year 2006 marked a turning point for industrial mathematics in Spain, with the launch of the project *Ingenio Mathematica*, known as i-MATH (http://www.i-math.org/). i-MATH was a CONSOLIDER singular research project for the period 2006–2012 in which a total of 350 research groups took part. It aimed to develop a complete activity research programme for Spanish mathematics, with the purpose of promoting and implementing strategic actions to increase – both qualitatively and quantitatively – Spain's mathematical presence at national and international levels.

In order to take this step, i-MATH has promoted several initiatives with a view to determining the current state of Spanish mathematics. These initiatives have provided a broad vision of current strengths and weaknesses, highlighting the need to act upon those areas in which mathematical community is not reaching its full potential, according to current economic development in the country. Furthermore, it has facilitated new progress in those areas and subjects where it already holds a significant or well-consolidated international position.

P. Quintela, A.B. Fernández, A. Martínez, G. Parente, M.T. Sánchez, *TransMath. Innovative Solutions from Mathematical Technology*
DOI 10.1007/978-88-470-2406-9_1, Springer-Verlag Italia 2012

Two of these initiatives are put forward in this book. Firstly, Chap. 2 presents the main conclusions derived from the TransMATH Demand Map; a national prospectus on the level of knowledge, use, and demand for mathematical technology by industry. Secondly, Chap. 3 is devoted to the TransMATH Supply Map, which analyses the experience acquired in technology transfer to business and industrial sectors by the research groups belonging to the i-MATH project.

The purpose of mapping this technological supply and demand, and an outline of the subjects the maps cover, is described in the following article on the grant agreement that was signed with the Spanish Ministry of Science and Education to mark the start of the i-MATH project:

> The design and annual update, together with the validation by an independent external committee, of a map showing the interactions and connections (both existing and potential) between mathematical research and technology transfer to the business and industrial sectors. The map will pay special attention to the detection of gaps in emerging fields, to the strengthening of existing fields, and to the identification of latent scientific-technological opportunities to be developed.

1.2
Mathematical technology in demand

For the drawing up of the TransMATH Demand Map, which was developed by CESGA Node (see http://mathematica.nodo.cesga.es/) in collaboration with the i-MATH board of directors, a project was designed for detecting problems in industry for which mathematicians could provide complementary or fundamental tools to solve them, determine the demand for mathematical training, and define, where necessary, new research lines in mathematics aimed at solving these problems.

This project constitutes a highly ambitious, pioneering venture, unique in the field of mathematics, in which a survey has been carried out with around 6,716 companies of 10 employees or more, distributed throughout Spain and representing all industrial and business sectors. From this sample, a total of 5,176 companies from 10 activity sectors have been selected for this book, excluding the sectors where the use of mathematical technology is not intensive or there is no specific supply from the research groups considered. To achieve this, a panel of experts from the academic, business, and industrial sectors has provided their advice and expertise. Specialists in CAD, numerical simulation, statistics, operations research and other fields of mathematics have participated in this panel, all with experience in the transfer of technology to companies. In addition to experts in technological consultancy at CESGA Node, their counterparts in consultancy at the Department of Statistics and Operations Research from the Universidade de Santiago de Compostela have also taken part in the processing of acquired information for the most recent version of the Map.

- In charge of the map:
  - Peregrina Quintela. Nodo CESGA Coordinator. Professor of Applied Mathematics. Universidade de Santiago de Compostela;
  - Andrés Gómez. Applications and Projects Administrator. Centro de Supercomputación de Galicia;
  - Wenceslao González. Professor of Statistics and Operations Research. Universidade de Santiago de Compostela.

- Panel of experts:
  - Aureli Alabert. Lecturer in Statistics and Operations Research at the Universidad Autónoma de Barcelona. Director of the Mathematics Consultancy Service. Member of the i-MATH Consulting Platform;
  - Alfredo Bermúdez de Castro. Professor of Applied Mathematics. Universidade de Santiago de Compostela. Coordinator of the ANEP Technology Transfer Area. Member of the i-MATH Consulting Platform;
  - Javier Bullón. Director General of FerroAtlántica R+D;
  - Laureano F. Escudero. Professor of Statistics and Operations Research. Universidad Rey Juan Carlos. Member of the i-MATH Consulting Platform;
  - Jesús Gil. Partner Director of INDIZEN Technologies;
  - Andrés Gómez. Applications and Projects Administrator. Centro de Supercomputación de Galicia;
  - Felix Martínez. Researcher in the Area of Mechanical Engineering. Centro Tecnológico Ikerlan;
  - Peregrina Quintela. Professor of Applied Mathematics. Universidade de Santiago de Compostela. Coordinator of the i-MATH Consulting Platform Committee;
  - Julio Rubio. Professor of Computation Science and Artificial Intelligence; Universidad de La Rioja. Member of the i-MATH Consulting Platform;
  - Arturo Soto. Departament of Calculus and Security. Grupo Antolín – Ingeniería.

- Technical team:
  - Mª Teresa Alonso. i-MATH Senior Technician of the Nodo CESGA;
  - Miguel Bermejo. EOSA Estrategia y Organización S.A.;
  - Miguel Costas. Head of field work team and consultant editor of the EOSA Estrategia y Organización S.A.;
  - Mª José Ginzo. Technician for the Department of Statistics and Operations Research at the Universidade de Santiago de Compostela;
  - Mónica López. Technician for the Department of Statistics and Operations Research at the Universidade de Santiago de Compostela;
  - Giuseppe Viglialoro. i-MATH Senior Technician of Nodo CESGA;
  - Jeff Palmer. Assistant in the preparation of the English version of the manuscript, Universidad Politécnica de Cataluña.

1.3
Mathematical technology on offer

To carry out the TransMATH Supply Map, a survey was developed to identify the starting point and evolution for each group taking part in the i-MATH project. The objective was to identify i-MATH groups with the capacity to transfer knowledge as well as proven experience in this field. It gave us a picture of their background in developing industrial projects, their interest in providing advanced technical training, and the range of mathematical services they offer.

The first edition of the TransMATH Supply Map was published in September 2007 on the i-MATH project webpage (see http://www.i-math.org/mapa_consulting/). Consequently, the map was updated in 2008, 2009, 2010 and 2011, and made available on the i-MATH website. All versions are currently available on the Spanish Network for Mathematics and Industry website, math-in$^{.net}$ (http://www.math-in.net/), which includes technical information about the activity of each research group. This information has been shared with companies, technological parks, governments and research groups linked to the R&D world.

The results included in Chap. 3 correspond to the sixth edition of the TransMATH Supply Map, and contain updated information about capacity, resources, and technological provision of a broad representation of Spanish mathematical research groups with a more consolidated background in transfer of mathematical technology. The post-processing of data provided by the research groups has been carried out by technicians at CESGA Node.

- In charge of the map:
    - Peregrina Quintela. Nodo CESGA Coordinator. Professor of Applied Mathematics. Universidade de Santiago de Compostela.

- Technical team:
    - Adela Martínez. i-MATH Senior Technician at Nodo CESGA;
    - Guadalupe Parente. i-MATH Senior Technician at Nodo CESGA;
    - Mª Teresa Sánchez. i-MATH Senior Technician at Nodo CESGA.

1.4
Economic activities involved in mathematical transfer

In order to facilitate the matching up of supply with demand for mathematical technology, the most noteworthy results of both the TransMATH Demand Map and the TransMATH Supply Map are broken down into sectors of economic activity. Table 1.1 shows the eleven economic activity groups which have been considered in this book.

These sectors were extracted from the European Industrial Activity Classification (NACE Rev.2), in accordance with the foreseeable applicability of mathematical

Table 1.1 Codes for the eleven economic activity groups considered in the results presented in this book, main sectors of economic activities which are included in each code, and corresponding NACE codes

Code	Main economic sectors included	NACE
Biomedicine & Health	Biomedicine and health	M75, Q86-Q88
Construction	Construction	F41-F43
Economics & Finance	Economics and finance	K64-K66
Energy & Environment	Energy, chemical and environmental	B05-B09, C19-C23, C32-C33, D35, E36-E39
Food	Food and clothing	C10-C15
ICT	Information and Communication Technology	J58-J63
Logistics & Transport	Logistics, transport and storage	H49-H53
Management & Tourism	Management services, social studies, tourism and entertainment	M69-M70, N77-N82
Metal & Machinery	Aeronautics, automotive and naval industries, metal and machinery	C24-C30
Public Administration	Public administration (only for mathematical technology on supply)	O84
Technical Services	Technical services (only for demand for mathematical technology)	M71-M74

techniques. The NACE codes for the economic activities included in each selected sector can also be seen in Table 1.1.

Both the demand for and supply of mathematical technology are mainly focused on economic activities included in these eleven selected sectors. However, the *Public Administration* sector was excluded from the TransMATH Demand Map since their economic activities are not usually included in business studies, whereas the *Technical Services* sector was ruled out of the TransMATH Supply Map due to its cross-sectoral nature, with corresponding services being redistributed into other sectors.

References

Eurostat – European Commission (2008) Nace Rev. 2 – Statistical classification of economic activities in the European Community. Office for Official Publications of the European Communities, Luxembourg

Quintela P, González W, Alonso MT, Ginzo MJ, López M (2010) TransMATH Demand: i-MATH map of company demand for mathematical technology. Nino-Centro de Impresión Digital, Santiago de Compostela

Quintela P, Sánchez MT, Martínez A, Parente G (2011) TransMATH: Investigadores en matemáticas para dar soluciones innovadoras. Project Ingenio Mathematica (i-MATH). http://www.i-math.org/files/File/documentos/mapa_consulting.pdf. Accessed 13 April 2011

i-MATH Map of Demand for Mathematical Technology (TransMATH Demand Map)

2

The TransMATH Demand Map analyses the results of a survey of 5,176 Spanish firms belonging to ten different sectors of economic activity. This chapter presents the main conclusions of the study in relation to the level of knowledge and use of mathematical technology by Spanish companies (such as computer-aided design, computer-aided engineering, statistical and operations research tools, and other mathematical techniques applicable to industry), the presence of personnel qualified in mathematics or statistics on their workforces, their requirements for mathematical services or qualified mathematicians or statisticians, and their interest in possible collaborations with universities or research centres.

2.1
Objectives and methodology

2.1.1
Objectives

The main purpose of this study is to detect the requirements and problems in different Spanish business sectors in which mathematics can act as either a fundamental or complementary tool to detect new possibilities for cooperation between mathematicians and industry, to define new research lines aimed at solving these problems, and to determine the demand for mathematical training. More specifically, the objectives of the TransMATH Demand Map are to:

- determine company needs and requirements concerning the incorporation of qualified professionals into the field of mathematics;
- detect training requirements in the fields of mathematics, statistics and operations research;
- identify lines of research of interest to companies in these fields;
- explore more deeply the different uses and applications that can be made available for representative sectors of the Spanish economy;
- detect current barriers to the adoption of these techniques by companies;

P. Quintela, A.B. Fernández, A. Martínez, G. Parente, M.T. Sánchez, *TransMath. Innovative Solutions from Mathematical Technology*
DOI 10.1007/978-88-470-2406-9_2, Springer-Verlag Italia 2012

- identify the predisposition and opportunities for collaboration between companies and universities and research centres;
- make known the benefits of mathematical techniques for small, medium-sized, and large companies;
- make known to the mathematical community the needs of the business sector in this field.

2.1.2
Methodology: population, sample and questionnaire

In order to carry out the study, a telephone survey of Spanish companies was conducted between March 24th and July 30th 2009. Geographically speaking, the survey covered the entire Spanish national territory.

The panel of experts which collaborated in the TransMATH Demand Map (see Sect. 1.1) carried out both the selection of the study population and the sample range, as well as the design of the questionnaire. Originally, 13 sectors of activity were selected for conducting the analysis of company demand for mathematical technology, discarding some significant economic areas which correspond to activities focused on education or to activities that are not usually covered in business studies. Of those 13 sectors surveyed, this book presents the results for ten of them (see Table 1.1 in Sect. 1.3), omitting the data corresponding to three sectors related to areas where either mathematical technology transfer is unusual, or mathematical research groups have still not developed specific supply: *Commerce*, *Timber & Paper* and *Miscellaneous Services* were the sectors omitted.

Subsequently, although 6,716 Spanish firms were surveyed, only the responses of 5,176 companies are analysed in this manuscript. The complete study can be consulted at the i-MATH project website, http://www.i-math.org/mapa_demanda/, or http://www.math-in.net/.

2.1.2.1
Study population

The Central Companies Directory (CCD) of the National Statistics Institute (INE – Instituto Nacional de Estadística) was adopted as the basis of the study population, updated as of January 1st 2008. However, the initial range was restricted to:

- firms belonging to ten of the eleven sectors of economic activity introduced in Table 1.1 (see Sect. 1.3), those being, *Biomedicine & Health*, *Construction*, *Economics & Finance*, *Energy & Environment*, *Food*, *Information & Communication Technology (ICT)*, *Logistics & Transport*, *Management & Tourism*, *Metal & Machinery* and *Technical Services* (*Public Administration* is excluded from this study);
- companies employing more than 10 people. Three groups depending on the number of employees are considered: the first corresponds to companies employing

Table 2.1 Distribution of population by sector and company size (number of employees). Source: Central Companies Directory of the National Statistics Institute (INE), January 2008

Sector\Number of employees	10–49	50–199	>199	Total
Biomedicine & Health	5,234	1,199	465	6,898
Construction	42,197	4,415	733	47,345
Economics & Finance	1,085	273	241	1,599
Energy & Environment	9,278	1,884	540	11,702
Food	9,001	1,328	330	10,659
ICT	3,305	717	296	4,318
Logistics & Transport	9,144	1,163	293	10,600
Management & Tourism	12,185	1,897	921	15,003
Metal & Machinery	12,000	1,948	559	14,507
Technical Services	4,816	655	239	5,710
Total	108,245	15,479	4,617	128,341

between 10 and 49 employees; the second, between 50 and 199 employees; and the third, with more than 199 employees.

Hence, the total number of firms making up the study population described amounts to 128,341. Their distribution according to sector of economic activity and number of employees is shown in Table 2.1.

2.1.2.2
Sample range

With regards to the statistical purpose of the data, the survey was designed to allow for analysis of the sample according to sector of economic activity and size of company, aspects of particular interest to this study. The size of the sample was 5,176 companies, and the type of sample employed was random sampling, with segmentation by firm size and sector of economic activity as shown in Table 2.2 which presents the sample distribution.

Regarding company size, allocation of the sample was not performed in a proportional way among the three employee strata. This was due, in the first instance, to similar levels of error being sought, which would not be possible using proportional allocation due to the large differences among the three employee strata sizes in the study population. In the second, the category of larger sized companies was considered to be of great interest for the project, since these firms often have more resources devoted to collaborative R&D projects with universities and research centres, and are usually where demand is highest for technology transfer. Using proportional sampling, larger firms would become significantly less represented in the sample range, and the study could have lost valuable information. Table 2.3 shows the distribution of the sample and study population according to number of employees. One observes that only 16% of companies in the considered study population

Table 2.2 Distribution of sample range by sector and company size (number of employees)

Sector\Number of employees	10–49	50–199	>199	Total
Biomedicine & Health	191	130	47	368
Construction	150	361	142	653
Economics & Finance	244	104	27	375
Energy & Environment	315	200	91	606
Food	248	146	53	447
ICT	159	127	53	339
Logistics & Transport	252	182	55	489
Management & Tourism	302	202	125	629
Metal & Machinery	331	190	121	642
Technical Services	409	150	69	628
Total	2,601	1,792	783	5,176

Table 2.3 Distribution of population and sample by company size (number of employees). Study population and sample sizes are indicated in brackets

Size	Population (128,341)	Sample (5,176)
10–49 employees	84%	50%
50–199 employees	12%	35%
>199 employees	4%	15%
Total	100%	100%

employ more than 50 people, whereas 50% of firms represented in the sample belong to the medium or large company categories.

Once the sample range for each company size stratum was established, allocation according to the ten selected strata of economic activity was performed, paying special attention to those groups previously selected by the panel of experts as being of greatest interest for mathematical technology transfer, such as *Biomedicine & Health, Economics & Finance, Energy & Environment, ICT* and *Technical Services* (see Table 2.4, where the distribution of sample range and study population by economic sector is presented). These sectors were oversampled since they are where we find the highest rate of demand for mathematical technology and the greatest quantity of users of mathematical and statistical tools; proportional allocation would reduce the presence of this category of company in the sample range.

In short, rather than basing the selection criteria for the sample size by stratum on optimality criteria (proportional sampling, minimum variance, etc.), the study population distributed by company size and sector was considered, alongside subjective criteria based on the transfer experience of the map's panel of experts in order to guarantee a reasonable representation of sectors and sizes of companies regarded as being the most strategic for this study. It is for that reason that in this TransMATH Demand Map a descriptive analysis of the sample data has been chosen, leaving the

Table 2.4 Distribution of population and sample (%) by sector. Study population and sample sizes are indicated in brackets

Sector	Population (128,341)	Sample (5,176)
Biomedicine & Health	5%	7%
Construction	37%	13%
Economics & Finance	1%	7%
Energy & Environment	9%	12%
Food	8%	9%
ICT	3%	7%
Logistics & Transport	8%	9%
Management & Tourism	12%	12%
Metal & Machinery	11%	12%
Technical Services	4%	12%
Total	100%	100%

analysis of its inference on the study population for later work, should this be necessary.

2.1.2.3
Questionnaire

The questionnaire employed was designed in accordance with the specific objectives of the survey. To facilitate its application, questions were arranged in four main groups:

1. characterisation of company;
2. knowledge and use of mathematical techniques:
   - computer-aided design (CAD) and numerical simulation, usually known as computer-aided engineering (CAE);
   - statistical and operations research tools (ST/OR);
   - other mathematical techniques applicable to industry (OMT);
3. needs and human resources in mathematical techniques;
4. collaboration and outsourcing with universities and research centres.

In order to facilitate data processing and statistical analysis, the survey database was stored and analysed with the Statistical Product and Service Solutions (SPSS) statistical package.

2.1.2.4
Further information

The questionnaire and detailed methodology related to this survey are available at the i-MATH project website (see http://www.i-math.org/mapa_demanda/ and http://www.math-in.net/), as well as in the technical document entitled *Informe Técnico del Mapa TransMATH Demanda* (*i-MATH Map of Demand for Mathematical Technology – Technical Report*).

2.2
Main results of the study

In this section, the statistical analysis carried out with the survey data is presented. Details of the results obtained according to company size (number of employees) and sector of economic activity are also included.

Firstly, since the choice of the sample was based on company size and sector, and a sample proportional to the real population was not considered, it was decided that a descriptive statistical analysis should be given to the sample data. This characterisation of companies is followed by a series of results concerning the mathematical techniques employed by these companies. Finally, a summary of company responses about needs and human resources in mathematical techniques, collaboration with universities or research centres, and consultancy requirements are also presented.

2.2.1
Characterisation of companies: by sector, company size and department

This section includes an analysis of the distribution of the sample according to sector and company size. In addition, the percentage of companies belonging to the sample which have departments for quality control, design or technology, R&D or new product development is shown. The latter is analysed according to whether it is located in Spain or overseas.

2.2.1.1
Distribution of firms by sector according to company size

It is worth repeating that sectors where economic activity is most closely related to mathematical technology transfer were oversampled, whereas sectors with a greater number of firms but previously minor interest were less represented in the sample (see Table 2.4, Sect. 2.1.2.2). Furthermore, in order to avoid giving excessive weight to small firms in the results, the proportion of medium and large-sized enterprises selected in the sample is higher than the proportion of these firms observed in the real population (see Table 2.3, Sect. 2.1.2.2).

Thus, Fig. 2.1 shows the distribution of sampled firms in each sector according to number of employees.

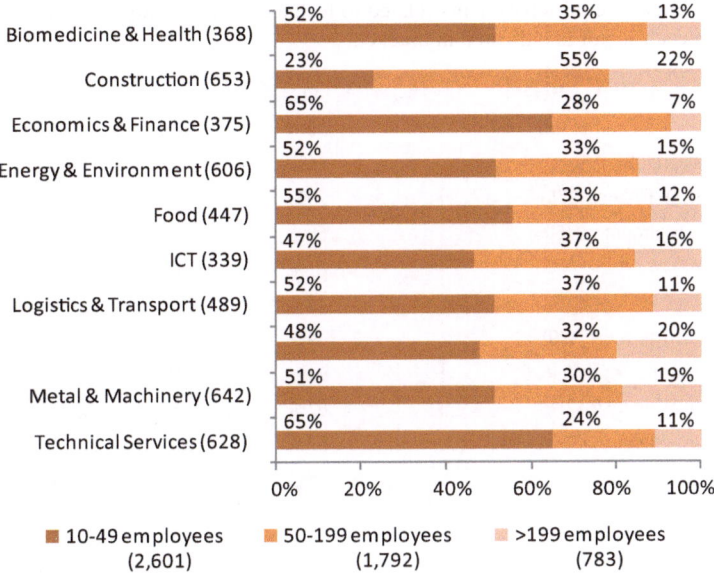

Fig. 2.1 Distribution of sample belonging to each sector by company size. The number of firms in each sector and the number of firms in each employee stratum are indicated in brackets

2.2.1.2
Existence of company departments related to technological transfer

In order to detect the existence of departments in firms which in theory might be more interested in mathematical technology transfer, one of the questions in the questionnaire referred to whether the company had any of the following departments:

- quality control department;
- design or technical department;
- R&D or new product development department.

From the responses, it follows that 37% of companies did not have any of these departments, whereas 52% had a quality control department, 34% a design or technical department, and 21% a R&D or new product development department.

Analysing this information by sector (see Table 2.5), more than 65% of companies whose activities involved *Energy & Environment*, *Food* and *Metal & Machinery* sectors had a quality control department. Moreover, in this latter sector, 60% of firms had a design or technical department, which is the highest percentage among all sectors. With regards to R&D or new product development departments, the most noteworthy are the *ICT* (39%), *Metal & Machinery* (32%) and *Energy & Environment* (31%) sectors. The lowest percentage corresponds to *Logistics & Transport*, where this department exists in only 7% of companies.

Table 2.5 Existence of departments related to technological transfer by sector. The number of firms in each sector is indicated in brackets

Sector	Quality	Design	R&D	None
Biomedicine & Health (368)	52%	14%	11%	45%
Construction (653)	46%	39%	10%	42%
Economics & Finance (375)	29%	19%	19%	59%
Energy & Environment (606)	68%	42%	31%	26%
Food (447)	67%	28%	27%	29%
ICT (339)	40%	44%	39%	29%
Logistics & Transport (489)	49%	11%	7%	49%
Management & Tourism (629)	41%	19%	11%	52%
Metal & Machinery (642)	67%	60%	32%	23%
Technical Services (628)	52%	46%	25%	28%
Total (5,176)	52%	34%	21%	37%

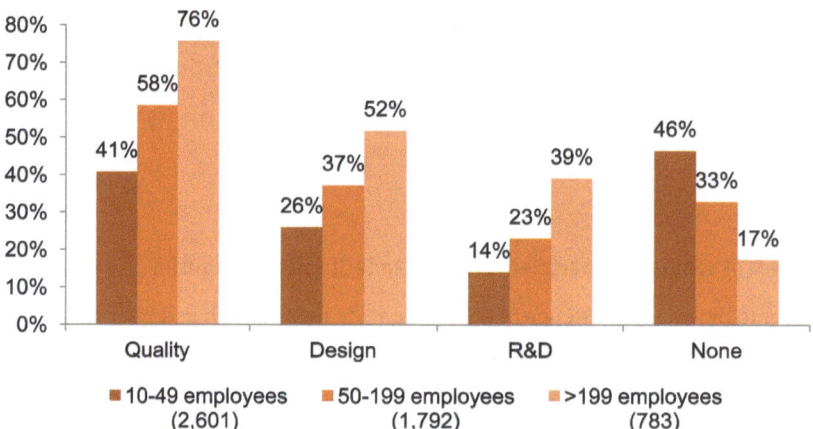

Fig. 2.2 Existence of departments related to technological transfer by company size. The number of firms in each employee stratum is indicated in brackets

Regarding company size, one observes that more than half of medium-sized companies in the sample and 76% of large companies possess quality control departments (see Fig. 2.2). Both design departments and R&D departments yield lower percentages in the three groups. In general, as company size increases, the number of firms with departments of any of the mentioned types also increases. On the other hand, this relation is inverted when those firms that have none of these departments are considered: the highest proportion (46%) corresponds to the smallest companies, while this percentage decreases as the number of employees increases, falling to 17% for large companies.

2.2.1.3
Location of R&D or new product development department

As mentioned above, among the different company departments, the R&D or new product development departments are of particular interest to this study (21% of those polled confirmed that this type of department exists in their firm). Special attention was paid to where these departments are situated, classifying the firms into three groups according to whether the R&D department is located in Spain, overseas or whether it is both overseas and in Spain. In regards to this question, the R&D department was exclusively in Spain in 86% of cases, while only 3% of companies in the survey stated that it was located overseas. Furthermore, 10% of firms said their R&D departments were both overseas and in Spain, and 1% of companies with this department did not answer the question (don't know/refusal).

In the analysis by sector, it is worth pointing out that 9% of firms involved in *Economics & Finance* had R&D or new product development departments overseas, while 16% stated that such departments were located both in Spain and overseas, yielding a result in this sector of 25% of companies with all or part of their R&D departments located overseas (see Fig. 2.3).

If the sample is divided by company size, as the number of employees increases, the percentage of companies which have all or part of the R&D or new product development department overseas also increases, ranging from 5% in small companies to 23% in the largest firms (see Fig. 2.4).

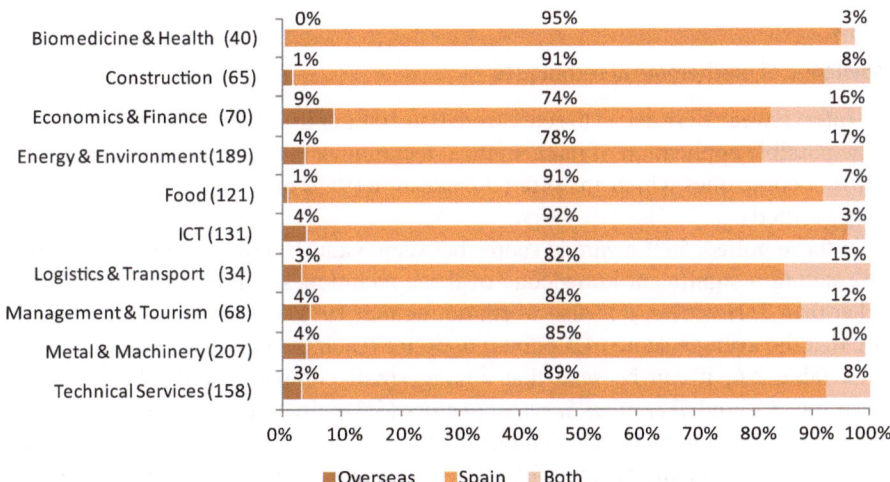

Fig. 2.3 Location of R&D or new product development department by sector. The number of firms which were asked by this question, that is, companies that had R&D or new product development department, is indicated for each sector in brackets. In this figure, 'No response' is not included for more clarification

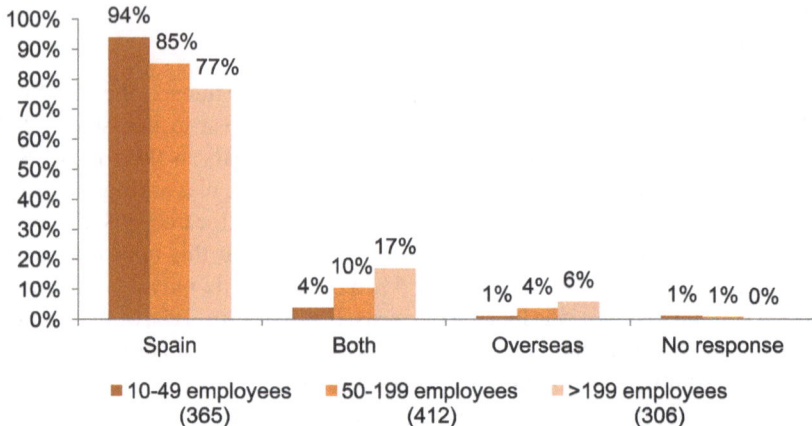

Fig. 2.4 Location of R&D or new product development department by company size. The number of firms which were asked this question, that is, companies that had R&D or new product development departments, is indicated for each employee stratum in brackets

2.2.2
Knowledge and use of mathematical tools: CAD/CAE, statistics and operations research, and other mathematical techniques

2.2.2.1
Knowledge of mathematical techniques

The interviewees were asked to score from 0 to 10 their level of knowledge as regards to the possible application of mathematical techniques in their companies. Here, the scores of companies which answered the question (95% of firms) are analysed (N/A being omitted).

Of those polled, 59% awarded scores between 5 and 10 points with, in particular, 30% stating a significant knowledge of these techniques with a rating equal to or higher than 7. However, it is also worth noting that 19% scored 0 points meaning in fact, that the average score (4.5 points) does not reach the pass mark of 5 points.

In Table 2.6, it can be seen that among small companies only those belonging to the *ICT* and *Technical Services* sectors (5.3 points) attain the pass mark for knowledge of mathematical techniques. Three sectors rise above 5 points in the medium-sized company group: *Metal & Machinery* (5.9 points), *Technical Services* (5.7 points), and *Energy & Environment* (5.1 points). On the other hand, it is in the large company group where the mean knowledge exceeds 5 points, being highly valued in the following sectors: *Metal & Machinery* (6.9 points), *Economics & Finance* (6.3 points), *Energy & Environment* (6.2 points), and *Technical Services* (6 points).

When one analyses the rating by sector, the greatest knowledge of the applications of mathematical techniques addressed in this study mostly correspond to large com-

Table 2.6 Mean level of knowledge of mathematical techniques by sector and company size (score from 0 to 10)

Sector\Number of employees	10–49	50–199	>199	Total
Biomedicine & Health	3.5	3.3	4.8	3.6
Construction	3.6	4.4	5.3	4.4
Economics & Finance	3.9	3.1	6.3	3.8
Energy & Environment	4.1	5.1	6.2	4.8
Food	3.7	4.7	5.5	4.3
ICT	5.3	4.8	5.0	5.1
Logistics & Transport	3.2	3.9	5.0	3.7
Management & Tourism	3.7	3.8	4.1	3.8
Metal & Machinery	4.6	5.9	6.9	5.4
Technical Services	5.3	5.7	6.0	5.5
Total	4.2	4.5	5.5	4.5

panies, followed by medium-sized companies. Furthermore, the scores of the latter are generally above those of smaller companies. Only in the *ICT* sector do small companies score higher (5.3 points) than the medium-sized companies (4.8 points), and large firms (5 points), as can be seen in Fig. 2.5.

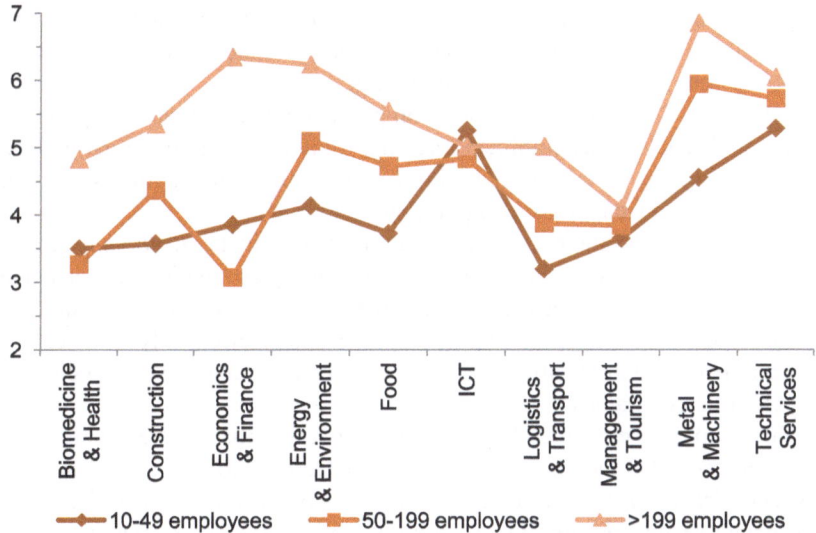

Fig. 2.5 Mean level of knowledge of mathematical techniques by sector and company size (score from 0 to 10)

2.2.2.2
Use of mathematical techniques: CAD, CAE, ST/OR and OMT

The following sections are devoted to analysing the use, on the part of Spanish companies, of the following mathematical tools:

- computer-aided design (CAD), and numerical simulation, usually known as computer-aided engineering (CAE);
- statistical and operations research tools (ST/OR), such as data analysis techniques or decision-making support techniques;
- other mathematical techniques (OMT), for example, geographical tracking systems, signal processing, computation, biomathematics, etc.

We present the percentages in use of mathematical techniques obtained, detailing in which specific areas they are mostly required. It is also indicated whether a company's mathematical techniques are conducted internally or externally. Following this, a special section compiles some results related to the type of software used in firms when mathematical techniques are applied internally.

Firstly, the interviewees were asked about the use of CAD techniques, that is, the use of computer-aided design for designing parts, plans, images or graphs. According to data gathered in this survey, 36% of those polled stated that these types of tools are applied in their companies. As regards to the application of computer calculation programs to simulate, predict or study the behaviour of products and/or processes (often known as CAE), such as thermal studies, mechanical forces or manufacturing processes, analysis of responses indicates that 14% of firms use CAE techniques.

The diffusion of statistical and operations research tools in the field of business was also studied. From the data obtained, one observes that almost half the companies (47%) stated that they use statistical techniques, data analysis techniques or decision-making support techniques in order to conduct, for example, customer or market analysis and quality control, as well as for planning, risk evaluation, logistics, resource assignation, and process optimization.

As far as OMT are concerned, 9% of companies confirmed that they applied mathematical tools such as geographical tracking, digital imaging or signal processing; geometry, design or visualization; bioinformatics or biomathematics; searching and codification of information, or computation.

Finally, let us highlight that among all the companies in the survey only 35% stated that they did not use any of the mathematical techniques mentioned. Consequently, 3,382 companies use one or more of the mathematical techniques included in the questionnaire. Nevertheless, of all the companies interviewed, only around 2% make use of mathematical applications in all areas covered in the survey, which amounts to a total of 84 companies.

If the sample is segmented by sector, one observes that the percentage of companies that use CAD increases considerably in *Metal & Machinery*, *Technical Services*, and *Construction*, with more than 50% (see Table 2.7). Next, the *Energy & Environment* sector appears (45%), giving slightly lower results but still above the mean. It can also be seen in Table 2.7 that among the 14% of sampled companies

Table 2.7 Percentages of companies using CAD, CAE, ST/OR, OMT or none of them by sector. The number of firms in each sector is indicated in brackets

Sector \ Mathematical tools	CAD	CAE	ST/OR	OMT	None
Biomedicine & Health (368)	13%	2%	47%	7%	47%
Construction (653)	54%	14%	33%	7%	36%
Economics & Finance (375)	19%	10%	48%	6%	43%
Energy & Environment (606)	45%	18%	53%	8%	28%
Food (447)	25%	12%	56%	4%	35%
ICT (339)	27%	8%	52%	16%	33%
Logistics & Transport (489)	10%	5%	50%	13%	44%
Management & Tourism (629)	17%	7%	44%	7%	47%
Metal & Machinery (642)	67%	29%	48%	7%	22%
Technical Services (628)	54%	21%	42%	16%	21%
Total (5,176)	36%	14%	47%	9%	35%

which use some type of CAE technique, the following sectors must be emphasised: *Metal & Machinery* (29%), *Technical Services* (21%), *Energy & Environment* (18%), and *Construction* (14%). The lowest value is found in those companies devoted to *Biomedicine & Health* (2%).

The most noteworthy sectors for ST/OR users are *Food* (56%), *Energy & Environment* (53%), *ICT* (52%) and *Logistics & Transport* (50%), since more than half the companies in these sectors applied statistical methods, data analysis techniques or decision-making support tools. On the contrary, the lowest percentage is found in *Construction* (33%).

For OMT, Table 2.7 shows that differences exist according to which sector the company belongs. Thus, the lowest percentages of OMT use (4%) are found in the *Food* sector, whereas the highest values correspond to *ICT* (16%), *Technical Services* (16%), and *Logistics & Transport* (13%).

When companies that did not use any of the techniques covered by the study are analysed, one finds that they belong to the following sectors: *Biomedicine & Health* (47%), *Management & Tourism* (47%), *Logistics & Transport* (44%), and *Economics & Finance* (43%). On the other hand, *Technical Services* and *Metal & Machinery* have the lowest percentage of firms that do not use mathematical tools (21% and 22%, respectively).

If the sample is divided according to the number of employees, the use of CAD (and CAE, respectively) techniques increases when the company size is larger, going from 30% (11%) in companies with fewest employees to 52% (22%) in the largest companies (see Fig. 2.6). Similarly, the percentage of firms that use statistical or operations research tools is higher in large and medium-size organisations (59% and 49%, respectively) than in small companies (41%). Moreover, as expected, the larger the firm the more they use OMT in their tasks, increasing from 7% for small companies to 14% for large companies. Finally, segmentation by company size reveals a

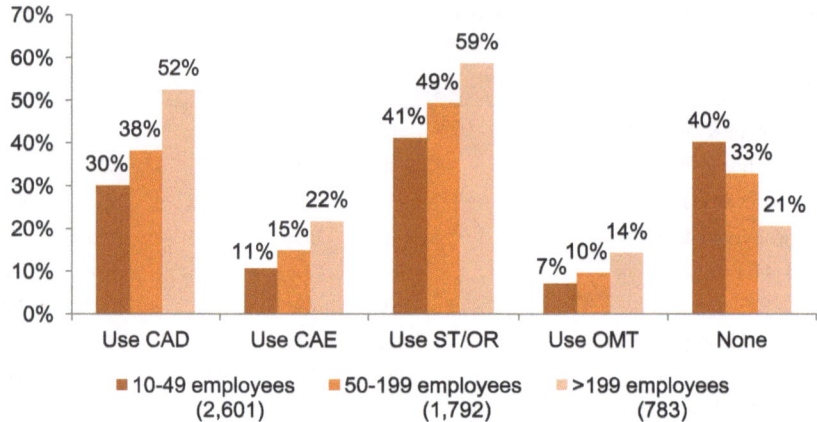

Fig. 2.6 Percentages of companies using CAD, CAE, ST/OR, OMT or none of them by company size. The number of firms in each employee stratum is indicated in brackets

significant increase in the use of mathematical techniques as company size increases, ranging from 60% in small companies to 79% in large companies (see Fig. 2.6).

2.2.2.3
Main areas of application of CAD/CAE, ST/OR and OMT

Next, we focus on the companies in the sample which stated that they use CAE, ST/OR or OMT, and we analyse in what areas these techniques are predominately applied in Spanish firms.

With regards to the purpose for which these companies use CAE, the interviewees were asked in which of the following areas they mainly apply CAE tools:

- mechanical or structural;
- thermal or thermodynamics;
- manufacturing processes: injection, printing, embossing, forging;
- electronics and/or electromagnetics;
- fluids: gases, liquids;
- acoustics or vibro-acoustics;
- environmental;
- other: for example, multiphysics.

In Fig. 2.7, one can observe that 53% of replies fall within the mechanical or structural fields and 37% in manufacturing processes. Note that, given the fact that the question required a multiple response, a company may have indicated one or several of the purposes and areas mentioned above.

The companies which stated that they used ST/OT tools also answered a question relating to the type of subjects or areas where statistical techniques, data analysis

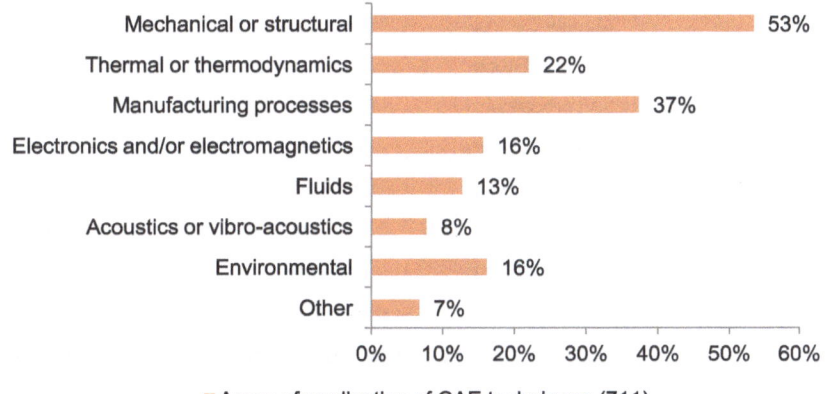

Fig. 2.7 Main areas where CAE techniques are applied by companies. The number of firms which were asked this question, that is, companies that used CAE tools, is indicated in brackets

techniques or decision-making support techniques applied. The possible answers to this question are detailed below:

- quality control;
- stock: control and optimization of stock;
- production processes: control and optimization of production processes;
- risk & financial analysis: analysis of risk, analysis of financial products;
- strategy, logistics & planning: strategy, decision-making, logistics, planning;
- customer/market/product studies: customer analysis, market studies, and product studies;
- exploitation of internal information: data mining, business intelligence;
- other: e.g. experimental design, clinical analysis.

Of all the companies using ST/OR in the course of their work, it transpires that 64% apply these tools to customer/market/product studies (see Fig. 2.8). A somewhat lower percentage, 50%, apply these tools to quality control, and 42% to strategy, logistics & planning. It should be noted that since the question admits multiple answers, the same company may indicate applications in several different areas or subjects.

A question in the questionnaire referred to what type of fields OMT were applied in companies using such techniques. The possible replies to this question were the following:

- digital imaging: graphs, video, animation, image recognition;
- geometric analysis: computational geometry, visualization, CAD development, symbolic methods;
- digital signal processing;
- GIS/GPS: design of geographical tracking systems such as GIS or GPS;

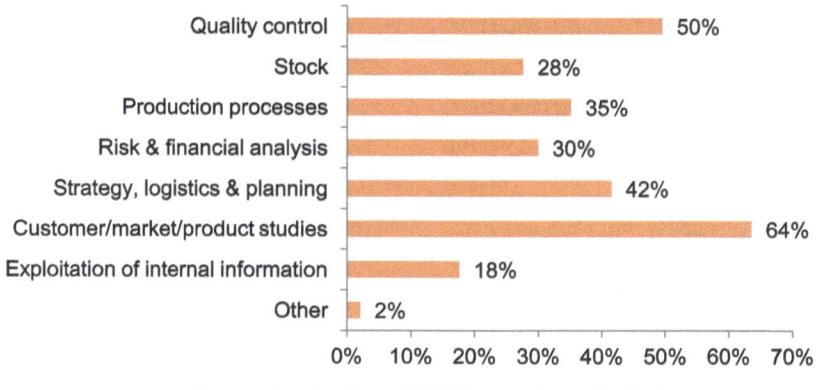

Fig. 2.8 Main areas where ST/OR techniques are applied by companies. The number of firms which were asked this question, that is, companies that used ST/OR tools, is indicated in brackets

- communication networks;
- codification of information: cryptography, computer security;
- computation: computation, computational algebra, language processing, numerical-symbolic algorithms;
- searching/processing of information: semantic web, algorithms for the internet;
- bioinformatics: genomics and proteomics;
- biomathematics: applications to life sciences and biomedicine, such as diagnostic techniques, medical prescriptions, drug administration, tumour growth and propagation, pest control, systems biology;
- other.

Of all the companies using OMT, 47% apply these tools to digital image processing (see Fig. 2.9). A slightly lower percentage (40%) stated that they were used in areas associated with design of geographical tracking systems such as GIS or GPS, while somewhat fewer but with practically the same percentage stated that they were applied to geometric analysis (25%) and communication networks (26%). Note that, since various answers were possible in reply, different areas of application may occur within the same company.

2.2.2.4
Internal/external use of CAE, ST/OR and OMT

In this section, companies using CAE, ST/OR and OMT are analysed, irrespective of whether mathematical techniques are employed internally, externally or both.

Figure 2.10 shows that 63% of companies in the survey use CAE on a solely internal basis, 15% of firms only use CAE externally, whereas 22% carry out CAE

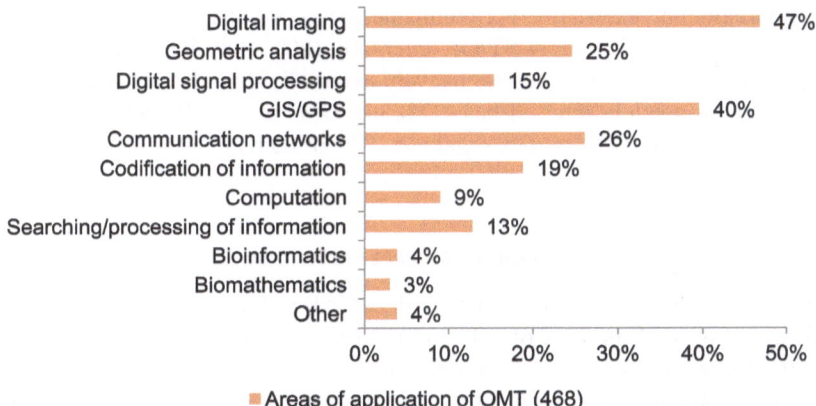

Fig. 2.9 Main areas where OMT are applied by companies. The number of firms which were asked this question, that is, companies that used OMT, is indicated in brackets

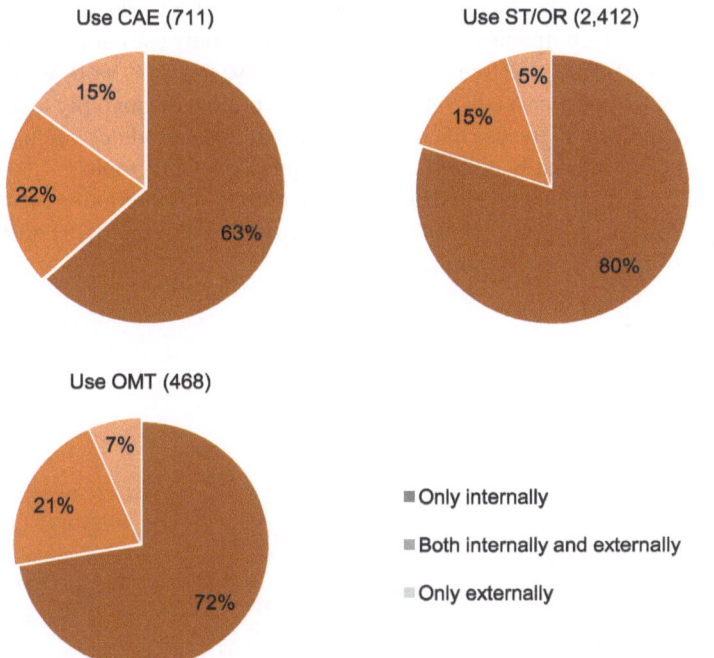

Fig. 2.10 Companies using CAE, ST/OR or OMT according to where mathematical techniques are carried out: internal, external or both. The number of firms which were asked this question, that is, companies that used CAE, ST/OR or OMT respectively, is indicated in brackets

both internally and externally. For companies using ST/OR tools, one may observe that 95% of companies apply ST/OR internally (80% only internally), while only 5% polled use such techniques exclusively externally. Finally, 72% of firms apply OMT tools on a solely internal basis, 7% use OMT externally, whereas OMT are conducted both internally and externally in 21% of companies.

2.2.2.5
Internal use of CAE and ST/OR: type of software used

In this section, those companies that conduct CAE or ST/OR internally are considered, and a study of the type of software used (commercial software, free software or both; customized software) is carried out.

Among companies that conduct CAE internally, 51% only use commercial software, whereas 15% of companies use free software on an exclusive basis (see Fig. 2.11). The remaining firms either use commercial and free packages simultaneously (23%) or declined to answer this question (11%).

If one deals with those companies that make use of ST/OR internally, Fig. 2.11 shows that 46% of companies using ST/OR internally only use commercial packages, whereas 19% of firms use free software exclusively, and 24% work with both types of packages. As in the case of CAE, 11% of those polled did not reply to the question.

Next, we will study the use of customized programs or modules among companies which carry out internally either CAE techniques or ST/OR tools.

Figure 2.12 shows that 42% of firms conducting CAE internally use customized programs for that purpose, while 70% of companies that employ ST/OR internally make use of customized modules.

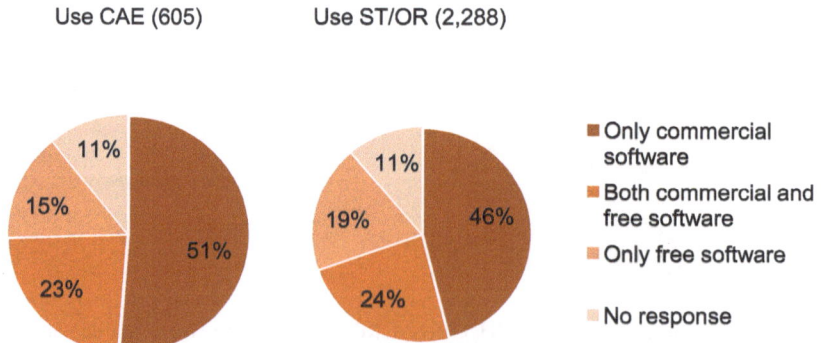

Fig. 2.11 Companies using CAE or ST/OR internally, according to the type of software used: commercial, free or both. The number of firms which were asked this question, that is, companies that used CAE or ST/OR internally respectively, is indicated in brackets

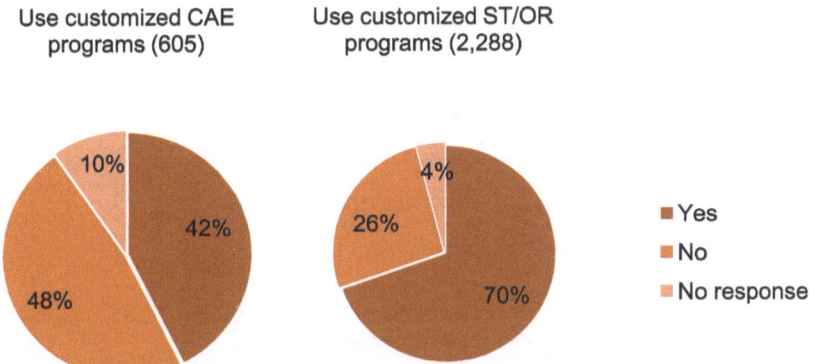

Fig. 2.12 Companies using CAE or ST/OR internally, according to their use of customized programs or modules. The number of firms which were asked this question, that is, companies that used CAE or ST/OR internally respectively, is indicated in brackets

2.2.3
Needs and human resources

2.2.3.1
CAD/CAE and ST/OR requirements

All the companies taking part in the survey were asked if they had any requirements regarding CAD/CAE or ST/OR; for example, if they needed information about or assessment of applicability of these techniques; training; implementation, development or improvement of programs. 4% of the sample replied that they required some assistance and 7% stated that they had some requirement in statistics, data analysis or decision-making support (ST/OR requirements). Thus in the whole survey, 535 firms reported the need for this kind of tool. Their distribution by sector can be found in Fig. 2.13.

From the calculation of the percentages for each sector, the CAD/CAE requirement increases in those sectors involving *Metal & Machinery* (7%, 42 firms), *Construction* (5%, 33 firms) and *Technical Services* (5%, 31 firms). For ST/OR techniques, *Economics & Finance* with 9% (34 firms) is the sector with the greatest requirement for these types of tools in terms of percentages, whereas *Management & Tourism* and *Technical Services* are the sectors with the highest demand in absolute terms with more than 40 firms each (around 7% of companies in these sectors).

As regards to company size, no appreciable differences are found in terms of a greater or lesser degree of need for information or assessment about subjects concerning CAD/CAE: around 4% of firms in each employee stratum (100 small companies, 67 medium-sized companies, and 31 large companies). If the comparison is now made for ST/OR needs, the percentage of firms with requirements increases

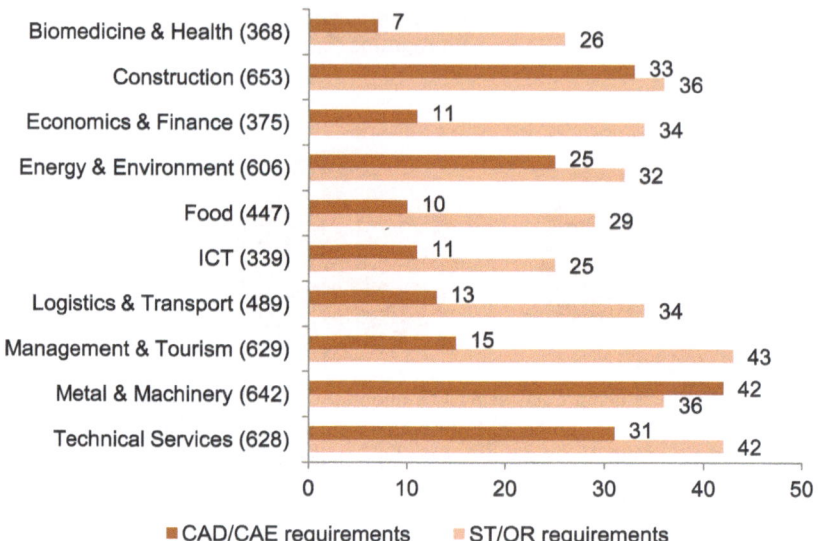

Fig. 2.13   Number of companies with CAD/CAE or ST/OR requirements by sector. The number of firms in each sector is indicated in brackets

slightly when the number of employees rises; going from 6% (150 firms) for small companies to 8% (62 firms) for large companies.

Type of CAD/CAE and ST/OR requirements

Those companies that stated their need for assistance in subjects relating to CAD/CAE were asked to classify their requirements into the following categories:

- information or assessment: information about or assessment of CAD/CAE applicability to the company;
- selection/implementation/validation of a CAD/CAE tool;
- training;
- definition or calculation of processes;
- customized development: customized development of CAD/CAE software or interfaces;
- integration of CAD with CAE or both into company processes;
- other.

Considering the 198 companies which stated a need for assistance with CAD/CAE as a basis, Fig. 2.14 shows how these companies are distributed according to their type of requirement. One may observe that 59% of such companies have requirements in information or assessment about the applicability of CAD/CAE to their company, and 53% in CAD/CAE training. Moreover, 21% of firms would be interested in selection, initial implementation and validation of a CAD/CAE solution adapted to

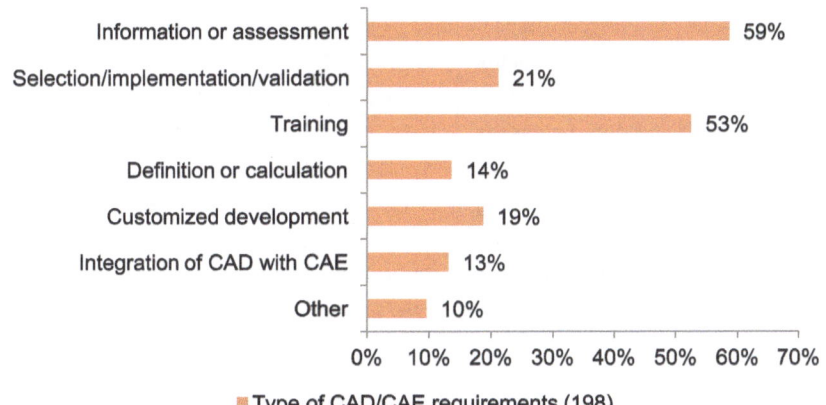

Fig. 2.14   Type of CAD/CAE requirements of companies that stated need for assistance. The number of firms which were asked this question, that is, companies that had any CAD/CAE requirement, is indicated in brackets

their needs, while 19% state requirements in customized development of CAD/CAE software.

The classification of ST/OR needs and requirements is the following:

- training;
- quality control;
- stock: control and optimization of stock;
- production processes: control and optimization of production processes;
- risk & financial analysis: risk analysis or financial products analysis;
- strategy, logistics & planning: strategy, decision-making, logistics and planning;
- customer/market/product studies: customer analysis, market studies, product studies;
- exploitation of internal information: data mining, business intelligence;
- other: for example, experimental design, clinical analysis.

Among the 337 companies which stated ST/OR requirements, the most frequently requested needs were found in the areas of customer analysis and market/product studies (41%), training (41%), quality control (32%), and strategy, decision-making, logistics and planning (29%), which in general coincide with the most widely used techniques, as seen in the previous section (see Fig. 2.15).

2.2.3.2
Human resources

The following sections are devoted to an analysis of whether or not the companies have personnel qualified in mathematics or statistics on their workforce, and what activities they are engaged in within the company. Analysis is also conducted into

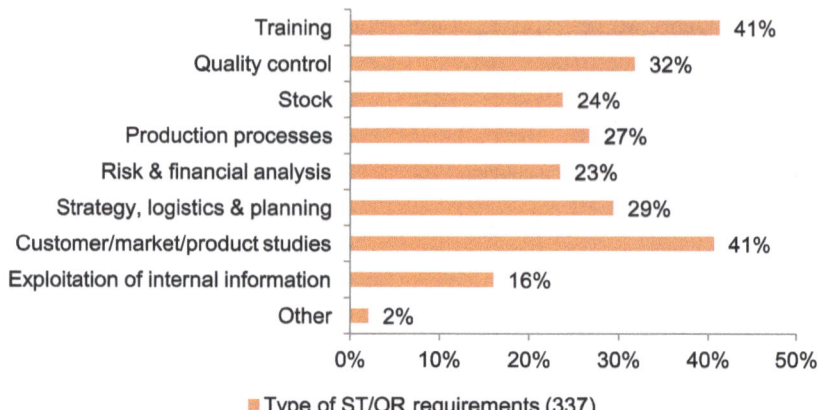

Fig. 2.15 Type of ST/OR requirements of companies that stated need for assistance. The number of firms which were asked this question, that is, companies that had any ST/OR requirement, is indicated in brackets

whether the companies require some kind of short or medium-term mathematical service or qualified personnel in order to apply any of the mathematical techniques addressed in this study.

Only 9% of companies stated that they have mathematicians or statisticians on their workforce, whereas 85% of firms did not employ personnel qualified in mathematics. The remaining interviewees (6%) did not reply to this question. This represents a total of 452 companies that employ mathematicians.

Figure 2.16 provides details of the percentage of companies employing mathematicians or statisticians by sector. The higher values correspond to the following sectors: *ICT* (21%), *Economics & Finance* (15%), and *Technical Services* (12%). As regards to company size, the larger companies more commonly have such employees (16%) as opposed to medium-sized companies or small companies (10% and 6%, respectively).

Furthermore, Fig. 2.17 provides details of the total number of mathematicians or statisticians by sector. Bear in mind that the total figure amounts to some 1,440 mathematicians spread over the 422 companies which answered this question (equalling 93% of firms stating they have mathematicians on their workforce).

The companies which replied they employed mathematicians or statisticians were asked about the fields in which these employees were engaged. The possible options were as follows:

- business informatics or systems;
- CAD/CAE;
- ST/OR: statistics, data analysis and decision-making support;
- OMT: other mathematical techniques;
- other tasks.

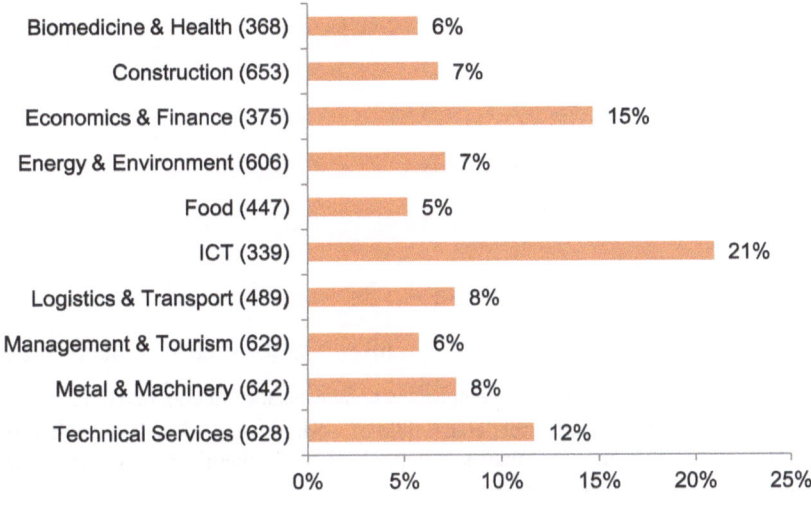

Fig. 2.16 Percentage of companies with mathematicians or statisticians on their workforce by sector. The number of firms in each sector is indicated in brackets

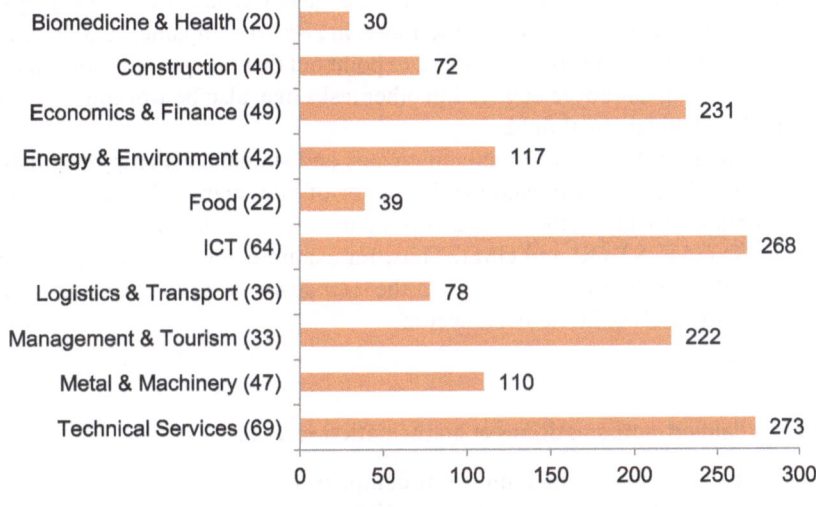

Fig. 2.17 Number of mathematicians or statisticians in those companies replying how many they employ by sector. The number of firms which answered this question in each sector is indicated in brackets

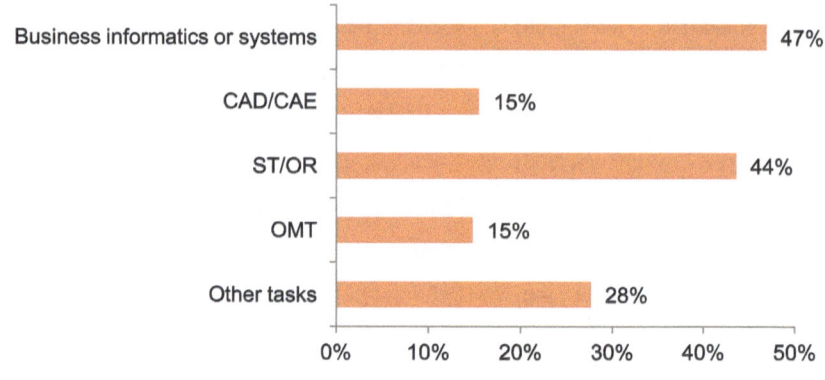

■ Fields in which mathematicians or statisticians are engaged (452)

Fig. 2.18 Mathematicians' or statisticians' field of work in companies. The number of firms which were asked this question, that is, companies with mathematicians or statisticians on their workforce, is indicated in brackets

Figure 2.18 shows from the analysis of the responses that mathematicians are engaged in fields associated with business informatics or business systems in 47% of companies, a percentage comparable to those engaged in statistics, data analysis or decision-making support (44%). Furthermore, in 28% of companies mathematicians are engaged in other tasks. It is important to point out that the question admitted multiple replies, so an employee engaged in other tasks may also be working at the same time in other mathematical fields.

Out of the sample of 1,440 mathematicians and statisticians employed by companies, Fig. 2.19 shows a breakdown by sector of how many of them are devoted, at least partially, to any of the tasks involving the main techniques addressed in this study: CAD/CAE, ST/OR and OMT. Thus, for example, 83% are engaged in fields directly related to their qualifications in the *Management & Tourism* sector, while this figure falls to 51% in the *Food* sector.

2.2.3.3
Need for qualified mathematicians or mathematical services

We now analyse short or medium-term company requirements for some types of mathematical services or the need to appoint qualified mathematicians or statisticians. Some 6% of interviewees stated that their company required mathematical services or needed to appoint personnel qualified in mathematics or statistics in order to apply one or more of the techniques considered in the survey; this represents a total of 286 companies from the 5,176 taking part in this study. 86% of firms, however, declared they did not have this type of need, and 8% of those polled decided not to answer.

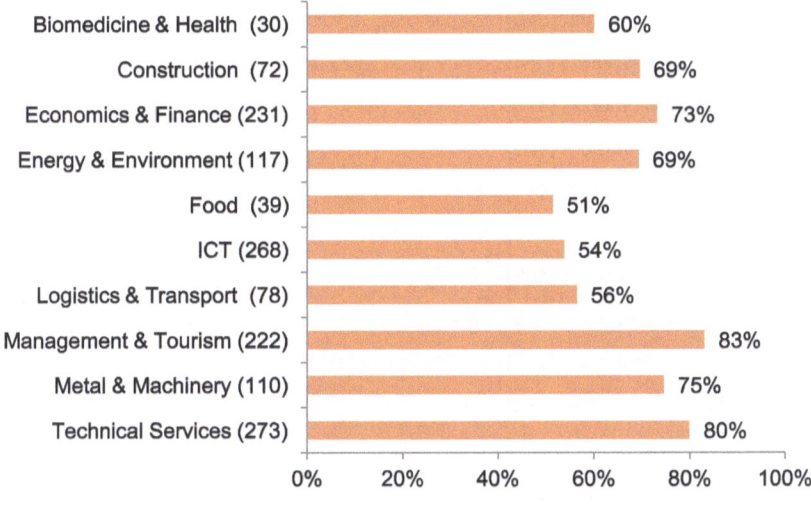

Fig. 2.19 Percentage of mathematicians or statisticians in the companies using CAD/CAE, ST/OR or OMT by sector. The number of mathematicians in each sector is indicated in brackets

Analysis by sector in Fig. 2.20 shows that the highest number of companies interested in appointing mathematical personnel corresponds to *Metal & Machinery* (41 firms, 6% of companies in the sector) and *Technical Services* (41 firms, 7% of companies in the sector). In terms of percentages, *ICT* (12%) and *Economics & Finance* (9%) are the sectors where the highest proportions of companies stated their need for some kind of service in this sphere.

When each group according to company size is compared, one observes that the largest companies give a figure of 8% (65 firms), which is 3% above the value for medium-sized and small companies (5% in both cases; 97 and 124 firms respectively).

2.2.4
Collaboration and outsourcing with universities and research centres

This section is first devoted to studying the collaboration and outsourcing by companies to universities or research centres over the last five years in all spheres, not just in mathematics. These collaborations include both training and research projects, and technological services. An analysis is then made of the interest shown by companies in possible collaborations with universities or research centres related to the techniques addressed in this study.

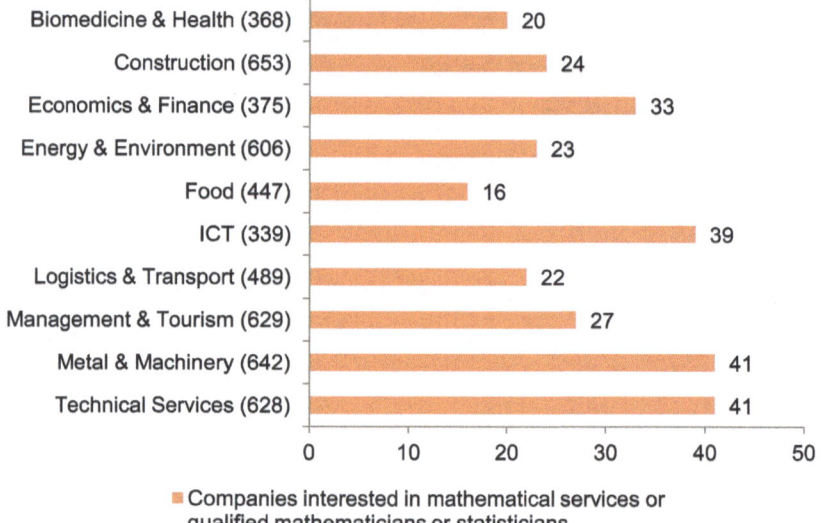

Fig. 2.20 Number of firms interested in mathematical services or qualified mathematicians by sector. The number of firms in each sector is indicated in brackets

2.2.4.1
Outsourcing and collaboration with universities or research centres in the last five years

What follows is an analysis of those companies that, between 2004 and 2009 (the survey was carried out in 2009), collaborated with universities and research centres in training, research or technological service projects, but not necessarily in fields related to mathematics. 33% of companies sampled responded affirmatively to the question, whereas 60% of those polled indicated that they had not collaborated with universities for the last five years (7% of interviewees did not reply to the related questions).

Sectors which have had the most collaboration or external contracts with universities or research centres over the last five years are *ICT*, and *Technical Services* (with 53% and 47%, respectively), while the sector with the fewest such collaborations is *Logistics & Transport* with only 19% of its companies outsourcing (see Fig. 2.21). Regarding company size, large companies have been those most involved in collaborations during the last five years, with a figure of 48% (see Fig. 2.22). Correspondingly, only 28% of small firms have recently collaborated with universities or research centres.

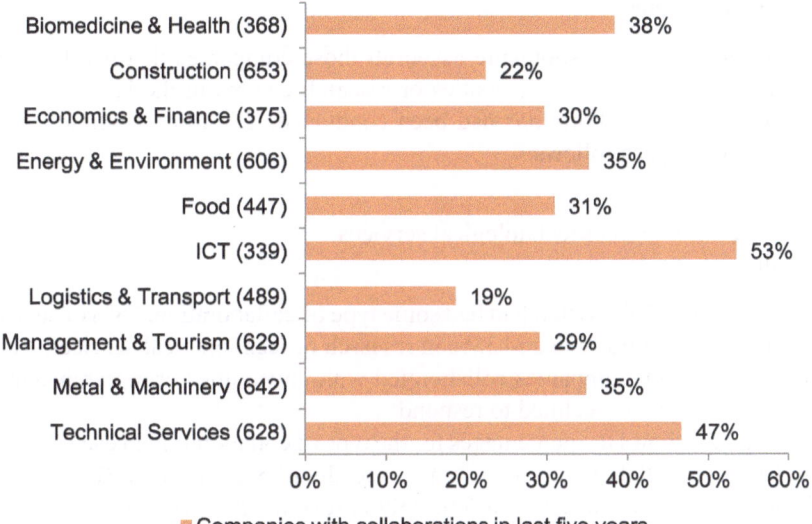

Fig. 2.21 Companies that have collaborated with universities or research centres in the last five years by sector. The number of firms in each sector is indicated in brackets

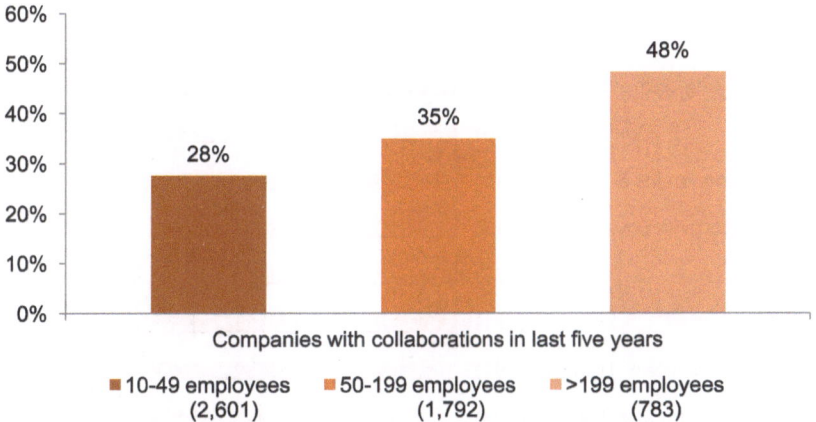

Fig. 2.22 Companies that have collaborated with universities or research centres in the last five years by company size. The number of firms in each employee stratum is indicated in brackets

Type of collaboration

One question in the questionnaire asked all those companies that had had collaborations or contracts with universities or research centres in the last five years, in what fields such collaborations had been conducted. The three possible replies to this question were as follows:

- training;
- services: research or technological services;
- both.

Based on the 1,721 firms that had had some type of collaboration, 78% of such cases were in the field of training and 38% in research or technological services. Furthermore, 18% of such companies collaborated with universities or research centres in both fields, while 3% declined to respond.

Figure 2.23 shows the percentages for the type of collaboration in each sector. The high degree of collaboration in training, exceeding 85%, in *Biomedicine & Health* and *Management & Tourism* (86% in both cases) is noteworthy. For collaboration on research or technological services levels are more modest, the highest percentages found in *Energy & Environment* (56%) and *Metal & Machinery* (52%). Among the sectors that have collaborated in both fields, it is worth highlighting *Energy & Environment* (23%), *Metal & Machinery* (22%), and *Food* (21%).

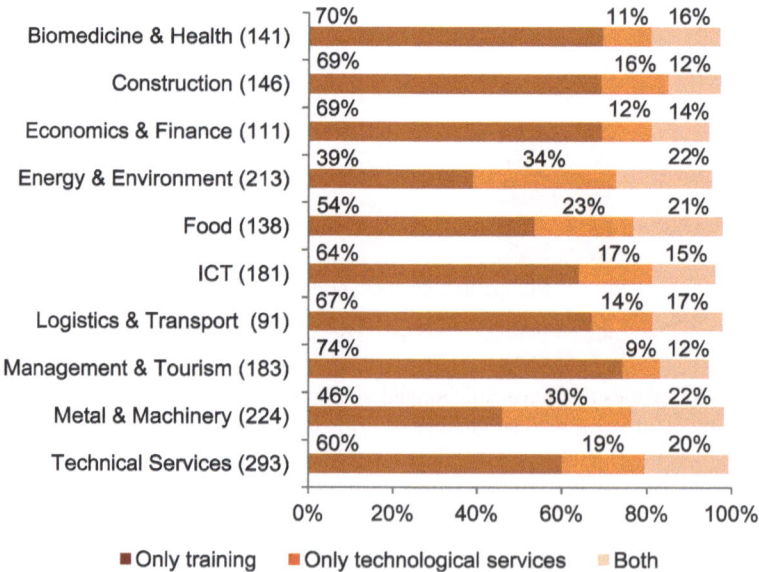

Fig. 2.23   Type of collaboration with universities or research centers by sector. The number of firms which were asked by this question, that is, companies that have had collaborations in the last five years, is indicated for each sector in brackets. In this figure, 'No response' is not included for more clarification

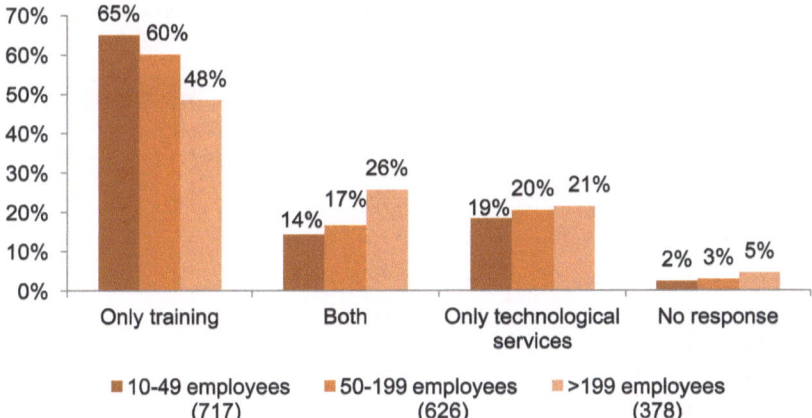

Fig. 2.24 Type of collaboration with universities or research centres by company size. The number of firms which were asked this question, that is, companies that have had collaborations in the last five years, is indicated for each employee stratum in brackets

On the other hand, small and medium-sized companies have collaborated most in training, with figures of 79% and 77%, respectively (see Fig. 2.24). As regards to research or technological services, large firms have had most collaboration or contracts of this type (47%), and they have also outsourced most in both fields (26%).

Degree of satisfaction

Companies collaborating with universities or research centres were asked about the degree of satisfaction according to points awarded from 0 to 10. Note that 7% failed to reply with a level of satisfaction.

One may highlight that only 2% of the companies which answered the question expressed a degree of satisfaction of less than 5, while 80% expressed a high degree of satisfaction with points equal to or higher than 7. Calculation of the average number of points awarded reveals that satisfaction with collaboration is high, with 7.5 out of 10.

In the analysis by sector (see Table 2.8 and Fig. 2.25), the highest points correspond to *Biomedicine & Health*, with 7.9. Nevertheless, no appreciable differences exist between sectors, since all values are higher than 7 points. Division according to company size reveals no appreciable differences from one group to another (except in the *Logistics & Transport* sector, where rating goes from 6.8 points for large firms to 8.0 points for small firms), although large companies returned slightly lower ratings.

Table 2.8 Mean level of satisfaction in collaborations with universities or research centres by sector and company size (score from 0 to 10)

Sector \ Number of employees	10–49	50–199	>199	Total
Biomedicine & Health	7.9	8.0	7.5	7.9
Construction	7.7	7.6	7.4	7.6
Economics & Finance	7.5	7.6	7.5	7.5
Energy & Environment	7.6	7.5	7.6	7.5
Food	7.2	7.7	7.3	7.4
ICT	7.5	7.5	7.0	7.4
Logistics & Transport	8.0	7.3	6.8	7.4
Management & Tourism	7.4	7.4	7.6	7.5
Metal & Machinery	7.0	7.2	7.1	7.1
Technical Services	7.5	7.4	7.5	7.5
Total	7.5	7.5	7.3	7.5

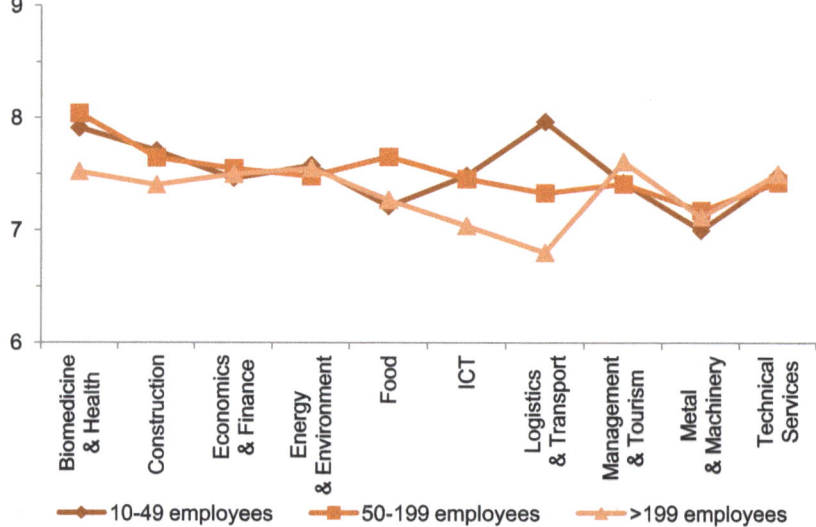

Fig. 2.25 Mean level of satisfaction in collaborations with universities or research centres by sector and company size (score from 0 to 10)

2.2.4.2
Interest in collaboration and outsourcing with universities or research centres

Regarding this issue, the questionnaire asked whether companies would be interested in collaborating with universities or research centres in end-of-year projects, Masters course projects or work experience placements within the sphere of mathematics. 33% of the total replies were affirmative, 44% were negative, and 23% corresponded to firms which did not answer the question. The high percentage of N/A responses is not particularly surprising, since most interviewees also indicated that they would need more information about the type and objectives of the collaboration in order to answer this question.

It can be seen in Fig. 2.26 that the sectors most willing to enter into collaboration, in terms of percentages, are *ICT* (48%, 163 firms) and *Technical Services* (45%, 282 firms), which account for almost half of these companies, followed by *Metal & Machinery* with 38% of companies (241 firms).

In the segmentation by company size (see Fig. 2.27), the largest firms expressed the most interest in collaborating with universities or research centres, with a figure of 44% (347 firms) of all those in the sample. The percentages corresponding to small and medium-sized companies were appreciably lower (30% and 32%, respectively), although there are more companies ready to collaborate in these employee strata in absolute terms (780 and 574 firms, respectively).

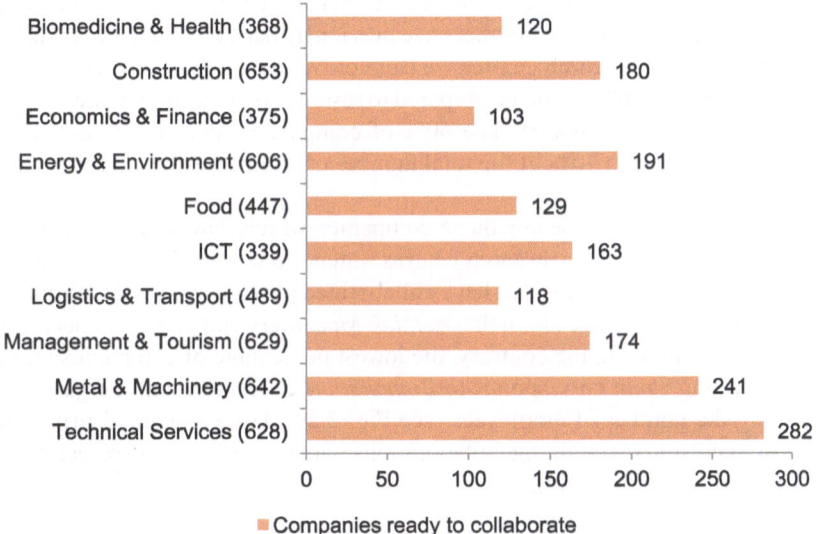

Fig. 2.26   Number of companies ready to collaborate with universities or research centres by sector. The number of firms in each sector is indicated in brackets

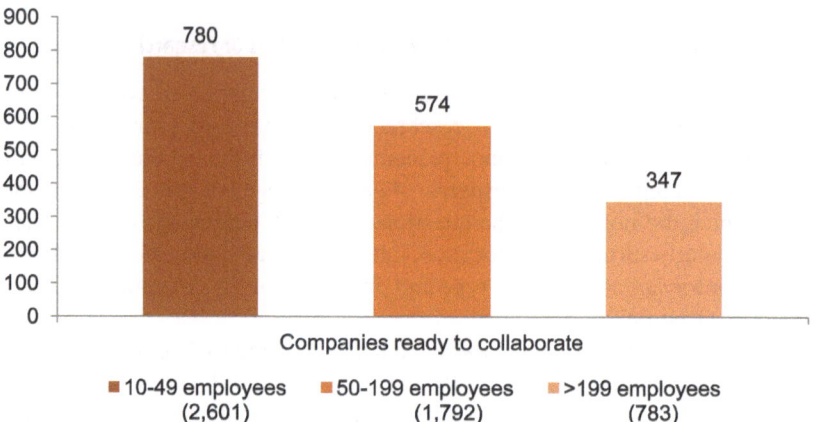

Fig. 2.27 Number of companies ready to collaborate with universities or research centres by company size. The number of firms in each employee stratum is indicated in brackets

Interest in maintaining collaboration

In this section we analyse those companies belonging to the group which replied that they have collaborated or had external contracts with universities or research centres in the last five years, and have expressed an interest in renewing that collaboration. Based on those 1,721 companies that have previously had some kind of collaboration with universities or research centres we can now reflect on their interest in doing so again, and find that 60% would be prepared to resume collaboration while 22% would not (18% declined to answer). The 60% of companies interested in collaborating again are equivalent to 20% of the total number of companies which participated in the study.

Figure 2.28 shows by sector those companies which, having previously collaborated with universities or research centres, intend to continue collaborating. In all sectors, the majority of firms that have collaborated previously stated their interest in collaborating again, above all, in the *Metal & Machinery* and *Technical Services* sectors (66% of firms). On the contrary, the lowest percentage of companies interested in collaborating again corresponds to *Biomedicine & Health*, with a figure of 50%. Regarding the number of employees (see Fig. 2.29), large and small firms showed the most interest in maintaining collaborations with universities and research centres (64% and 61%, respectively).

2.2.4.3
Interest in receiving a visit from mathematical technical staff

Those polled were asked about their interest in receiving a visit from a mathematical technical expert. This question was posed to the 864 companies stating their need for

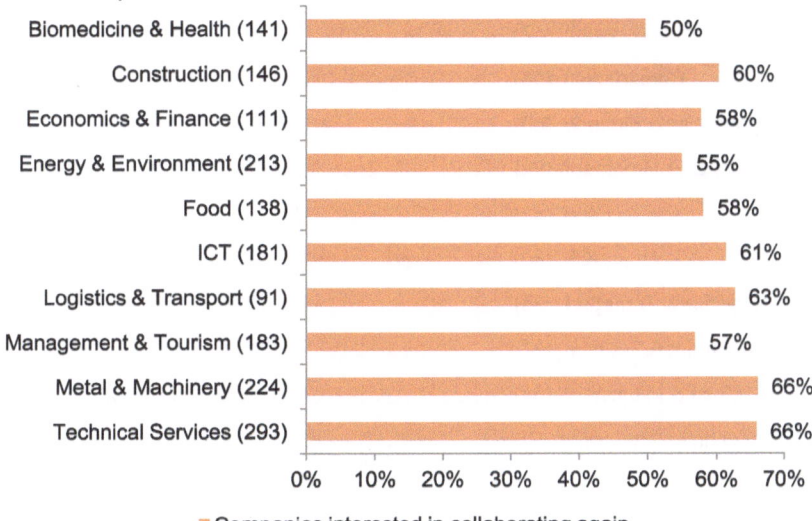

Fig. 2.28 Companies interested in collaborating again with universities or research centres by sector. The number of firms which were asked this question, that is, companies that have had collaborations in the last five years, is indicated for each sector in brackets

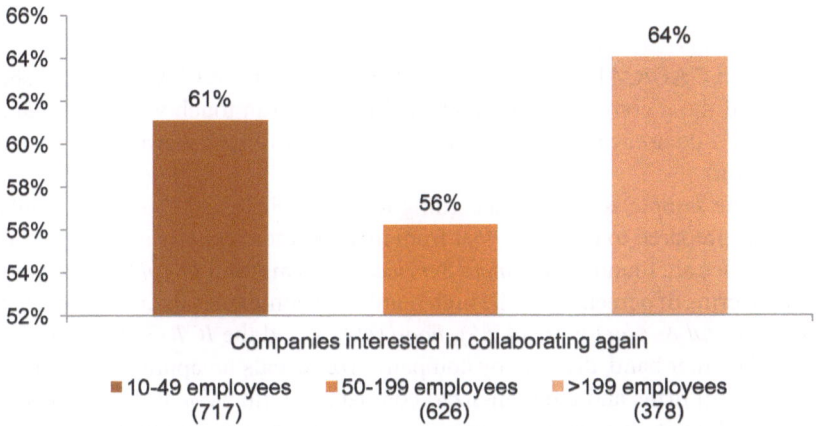

Fig. 2.29 Companies interested in collaborating again with universities or research centres by company size. The number of firms which were asked this question, that is, companies that have had collaborations in the last five years, is indicated for each employee stratum in brackets

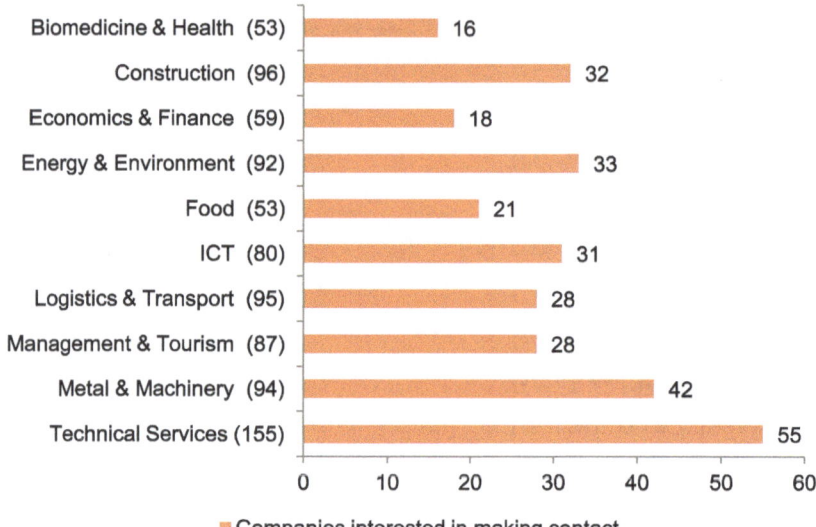

Fig. 2.30 Companies interested in making contact with mathematical technical experts by sector. The number of firms which were asked this question, that is, companies that had requirements in CAD/CAE or ST/OR, or applied OMT, is indicated for each sector in brackets

assistance in CAD/CAE or ST/OR as well as those that used OMT. Results showed that 35% of these companies are interested in getting in touch with technical staff, whereas 54% do not wish to receive such a visit (11% of interviewees did not answer this question).

When the sample is divided according to sector, Fig. 2.30 shows the number of companies prepared to receive a visit from mathematical experts in each sector. The highest values are found in *Technical Services* (55 firms) and *Metal & Machinery* (42 firms). In terms of percentages, the high number of companies interested in receiving a visit in *Metal & Machinery* (45%), *Food* (40%), and the *ICT* (39%) sectors stand out. On the other hand, division by company size reveals no appreciable differences between small firms and medium-sized companies, with 34% of those polled interested in such visits, while large companies seem to be just slightly more interested, with a figure of 37%.

2.3
Conclusions

This chapter contains the analysis of a survey of the business and corporate demand for mathematical technology in Spain, which has enabled us to:

- analyse the knowledge and use of mathematics in industry;
- detect the needs and problems in different Spanish business sectors where mathematical techniques could constitute a fundamental or complementary tool;
- attract the interest of companies with a view to some kind of collaboration with universities or research centres.

Next, we summarise the main conclusions about this study related to the level of implementation of mathematical techniques in firms, and the identification of primary company needs where there is a demand for mathematical technology and services.

2.3.1
Use of mathematical techniques: CAD, CAE, ST/OR and OMT

Firstly, let us highlight that amongst all the companies in the survey only 35% stated that they did not use any of the mathematical techniques mentioned, that is, 65% of those polled do actually use some mathematical tools in their companies. Taking into account the number of employees, those which do use mathematical tools range from 60% among small firms to around 80% in large companies.

In first place, with a figure of 47% of all companies surveyed, is the use of statistical tools, data analysis and decision-making support techniques (ST/OR), which find their highest degree of implementation in *Food* (56%), followed by *Energy & Environment* (53%), *ICT* (52%) and *Logistics & Transport* (50%). These techniques are used above all for customer analysis and market or product related studies.

Computer-aided design (CAD) comes second, since it is used in 36% of the companies polled, particularly in the *Metal & Machinery* sector (67%). Third position, and with considerably lower values, is occupied by computer-aided engineering (CAE) with 14% of firms, the greatest degree of implementation once again being in *Metal & Machinery* (29%). This tool is applied for conducting mechanical or structural calculations in approximately half of the cases (53%).

In last place with 9% is the use of other mathematical techniques (OMT), with almost double this percentage in the *Technical Services* and *ICT* sectors (16%). In almost half of the companies using OMT (47%), these techniques are used for the processing of digital images and in 40% of cases for the design of geographical location systems such as GIS or GPS.

For all of the previous techniques, the larger the company the more extensive is their use. Specifically, the use of CAD in large companies exceeds the use in small companies by 22%, and by 14% in medium-sized firms. Similar percentages are found for ST/OR, whereas the difference is narrower where CAE and the use of OMT are concerned. The effort that many small companies are making to incorporate these tools into the course of their work is clearly demonstrated by the corresponding percentages. In fact, 60% of small companies stated that they use some kind of mathematical technique.

2.3.2
Use of mathematical techniques: software

Approximately half of firms only use commercial software in order to internally apply CAE (51%) or ST/OR (46%) techniques. Correspondingly, 38% of those polled which are using CAE stated that they worked (either totally or partially) with free software, while 43% of declared ST/OR users applied internally these tools (either totally or partially) by means of free packages.

Moreover, 42% of firms applying CAE internally use customized programs/modules for that purpose. This percentage reaches 70% of companies in the case of ST/OR techniques.

2.3.3
Requirements in mathematical tools

Only 337 firms polled stated their need for assistance in statistics, data analysis or decision-making support. *Economics & Finance*, with 34 companies, is the sector with the greatest requirement in these types of tools. The greatest demand in this area is for training (41%), and for use in customer analysis and market or product studies (41%).

Requirements in CAD/CAE, however, are found in 198 companies, the highest values being in the *Metal & Machinery* sector (42 firms). As regards to the type of requirement, more than half of such firms (59%) stated their need for information or assessment about the applicability of CAD/CAE techniques in their companies.

2.3.4
Human resources

With regards to human resources, 452 firms (9% of companies in the study) indicated that they had mathematicians or statisticians on their staff. In almost half of the cases, these employees are engaged in business informatics or business systems (47%) or statistics, data analysis, and decision-making support (44%).

Furthermore, 286 companies (6% of those polled) state that they require mathematical services or personnel qualified in mathematics or statistics; the sectors with the highest demand being *ICT* with 12%, and *Economics & Finance* with 9%.

2.3.5
Collaboration and outsourcing

The percentage of companies expressing willingness to collaborate with universities or research centres is noteworthy (33% of those polled); the highest values corresponding to *ICT* (48%) and *Technical Services* (45%).

This percentage is similar to that of those companies that have already collaborated in the last five years (33% of those polled). Such collaboration mainly took place in areas of training, where the mean degree of satisfaction expressed by the companies involved was 7.5 points (scored from 0 to 10). This is probably the reason why more than half of such firms would be prepared to collaborate again with universities or research centres.

In general, companies in the study perceive that mathematical techniques are important. In fact, 35% of firms stating their need for assistance in CAD/CAE or ST/OR as well as their use of OMT would like to be contacted by mathematical technical experts.

2.3.6
Main conclusions

From the previous conclusions, we can suggest that there is a certain level of implementation of mathematical knowledge, and that companies are aware of this. For example, in large companies the mean knowledge exceeds 5 points out of 10, reaching the highest rating in *Metal & Machinery*, *Economics & Finance*, and *Energy & Environment*. Nevertheless, it is difficult to evaluate to what extent companies use of mathematical techniques goes beyond the purely instrumental (use of spreadsheets or standard statistical packages in their most basic functionality, for example).

Moreover, we believe that there remain many companies which are not aware of industrial applications of mathematics or the advantages that might arise by including mathematicians or statisticians on their staff. We believe that the reality of this situation is not fully appreciated by either the community of mathematical researchers or industry, where it is often thought that mathematics are far removed from the industrial and business worlds. However, mathematics is in fact becoming increasingly interdisciplinary, appearing in contexts in which it is difficult to be separated from other spheres of knowledge, and giving rise to industrial applications with greater frequency.

It is also very important to point out that around one fifth of companies in the survey stated that they had a R&D or new product development department. Such departments are highly correlated with technological transfer in the industry. Furthermore, a high percentage of companies are prepared to collaborate with universities and other research centres. Even more noteworthy is that most of these companies have previously participated in this type of activity in recent years. The conclusion seems clear: if companies collaborate with universities, and they are satisfied with the obtained results, then they wish to repeat the experience. Thus, further efforts must be made to overcome the obstacle posed by what appears to be a lack of knowledge in some companies about the possibility of participating in such collaboration. In fact, the upmost requirements of those polled are related to training and assessment of application of mathematical tools in their firms. To achieve this aim, dissertations, post-graduate projects and even doctoral theses carried out by new mathematicians can help to bring about closer relations with companies. Experience shows that this kind of timely relationship, linked to higher education and training, can lead to fruitful collaboration in the field of innovation and research.

Finally, let us remark that there still remains a long way to go in bringing mathematical technology to a greater number of firms, and, above all, that cases in which this has been successfully achieved should be made more widely known in order to increase company receptivity to these opportunities, since at present only a small number of companies are truly aware of its potential.

References

Quintela P, González W, Alonso MT, Ginzo MJ, López M (2009) Informe técnico del mapa TransMATH demanda. Project Ingenio Mathematica (i-MATH). http://www.i-math.org/mapa_demanda/Informe_tecnico.pdf. Accessed 11 May 2010

Quintela P, González W, Alonso MT, Ginzo MJ, López M (2010) TransMATH Demand: i-MATH map of company demand for mathematical technology. Nino-Centro de Impresión Digital, Santiago de Compostela

i-MATH Map of Supply of Mathematical Technology (TransMATH Supply Map)

3

This chapter presents the supply and expertise offered in mathematical technology transfer by the 62 research groups belonging to the i-MATH Consulting Platform in 2011. Firstly, a general characterisation of the groups and an overview of each of them (contact information, research lines, etc.) are given. Then the main areas of mathematics in which the research groups are currently working are listed. Finally, the supply of mathematical technology and the current expertise of the 62 groups are shown by sector. In particular, detailed information about consulting and training expertise offered by the research groups is presented classified by sector, as well as their expertise in software development and use of free and commercial software.

3.1
Objectives and methodology

3.1.1
Objectives

This study aims to identify capacity and expertise in the transfer of knowledge and technology to industry from Spanish mathematics research groups which belonged to the project *Ingenio Mathematica*. The main objective was to develop a map of capabilities that would integrate detailed information on the most representative groups working to create partnerships and interact with companies in different industrial sectors.

Overall, the TransMATH Supply Map aims to achieve several objectives, mainly directed at two types of stakeholders or target readers:

- A wide range of business and commercial firms, as well as non-profit R&D organisations, in order to present a complete document that puts forward new information on mathematical, statistical and operational research techniques and their application in the improvement of processes and products in industry. The map seeks to fill knowledge gaps in industry regarding areas of work in which collaboration with mathematical groups is possible. This lack of knowledge, identified

P. Quintela, A.B. Fernández, A. Martínez, G. Parente, M.T. Sánchez, *TransMath. Innovative Solutions from Mathematical Technology*
DOI 10.1007/978-88-470-2406-9_3, Springer-Verlag Italia 2012

in previous studies within an overall European context[1], has been evaluated from a Spanish perspective through a survey conducted to produce the TransMATH Demand Map, as well as tracking interaction with businesses during the development of the i-MATH project for the entire period 2007–2012. The set of actions carried out have shown that highlighting possible areas of collaboration to companies yields a strong corporate demand for mathematical technology to solve real industrial problems. The map aims to significantly increase collaborations between research groups and industry, solving new problems and meeting unresolved needs to strengthen the competitivity of companies.

- The mathematical community at large, and those researchers and scientists from other disciplines with an interest in mathematical knowledge transfer to industry in particular, can use the Map to view aggregate information about the research lines in which Spanish groups are currently working. In addition, from the data on industrial demand of mathematical techniques, it is easy to identify new areas of research oriented to the needs of business but not yet covered by Spanish mathematicians.

3.1.2
Analysis of research groups

The first edition of the TransMATH Supply Map was published in 2007. Since then, four new editions have been completed, with several updates in 2008, 2009, 2010 and 2011[2]. The survey used as a basis for the information contained in this chapter can be found in the complete study, at the i-MATH project website, http://www.i-math.org/mapa_oferta/, or http://www.math-in.net/. It was disseminated to all regions of Spain, among all i-MATH groups integrated in the Consulting Platform. This Platform was formed by the Spanish mathematical research groups with most experience in industrial mathematics and it aimed at increasing the use of mathematics in the manufacturing sector. During the life of the i-MATH project, intense activity to promote relationships between universities and companies has been developed. Through this intense activity a great number of collaborations with industry have emerged.

The annual increase in the number of research groups integrated into the Consulting Platform that took part in consecutive editions of the map is showed in Fig. 3.1. For the last two editions, the data collection has been performed using a web tool that allows research groups to easily incorporate their technological supply. Information about the last edition of the map, the use of the website, and deadline for submission was sent to all groups on July 22nd 2011. Since this date 62 responses have been re-

---

[1] See the main conclusions presented in the Final Conference of the Forward Look on Mathematics and Industry, convened by the European Science Foundation in Brussels on 2 December 2010, and related studies, referenced in the Forward Look website, www.ceremade.dauphine.fr/FLMI/.

[2] For the complete catalogue of services offered by Spanish researchers, see Quintela P, Parente G, Sánchez MT, Fernández AB (2012) Soluciones matemáticas para empresas innovadoras. Catálogo de servicios ofertados por investigadores españoles. McGraw-Hill, Madrid.

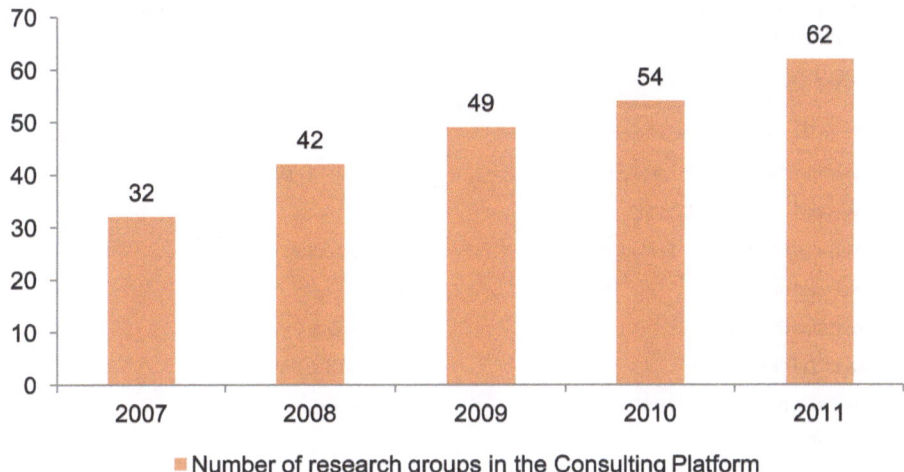

Fig. 3.1 Evolution of the number of research groups in the Consulting Platform from 2007 to 2011

ceived from the groups, out of which 8 correspond to new groups and the remaining 54 to updates.

Charts and tables contained in the following sections of this chapter include information about all 62 research groups integrated in the i-MATH Consulting Platform in 2011. All of them took part in the survey.

3.1.2.1
Characterisation of research groups: university or research centre

The 62 research groups, which participated in the Consulting Platform in 2011, belong to 31 Spanish universities or research centres as can be seen in Table 3.1.

In this table the acronyms of the research groups that will be used throughout this chapter are shown. More detailed descriptions of each group, such as contact information or main research lines, can be found in Sect. 3.2.

3.1.2.2
Characterisation of research groups: Spanish regions

Figure 3.2 shows the regional distribution of the 62 research groups. As the chart shows, there is a wide distribution throughout the regions, with representation of groups in all regions across the country.

The regions Galicia and Catalonia contain the highest number of groups (10 research groups); followed by Andalucía with 9; and Madrid and Valencia with 7 and 6 groups, respectively.

Table 3.1 Distribution of the 62 research groups in the Consulting Platform by university or research centre

University/research centre	Research groups (acronyms)
Instituto Nacional de Técnica Aeroespacial	ADF.
Universidad Autónoma de Barcelona	GSD, KINETIC, MCS-UAB.
Universidad Carlos III de Madrid	AALN.
Universidad Complutense de Madrid	ACEIA, modsol.
Universidad de Almería	TAPO.
Universidad de Cádiz	GAUCA, Orel.
Universidad de Cantabria	CAG, CYOPT, DATFUN.
Universidad de Castile-La Mancha	OEDgroup.
Universidad de La Laguna	GOMA.
Universidad de La Rioja	Psycotrip.
Universidad de las Islas Baleares	TAMI.
Universidad de Las Palmas de Gran Canaria	DDA, TD-ULPGC.
Universidad de Málaga	EDANYA.
Universidad de Murcia	GIO, GOR.
Universidad de Oviedo	GMFN.
Universidad de Sevilla	GIOPTIM, GPB97, LOGRO, M2S2M, TOREFA.
Universidad de Valencia	RUTYLO, RUTYMETA.
Universidad de Valladolid	GEUVA, GSO.
Universidad de Zaragoza	GIOS, MODESI.
Universidad del País Vasco – Euskal Herriko Unibertsitatea	EOPT, TTM.
Universidad Politécnica de Madrid	DEPREN, GSCUPM.
Universidad Politécnica de Valencia	Funaphy, HYPCHAOP, InterTech.
Universidad Pública de Navarra	DECYL, EE.
Universidad Rey Juan Carlos	RiTO.
Universidade da Coruña	M2NICA, MODES.
Universidade de Santiago de Compostela	EDnL, GRID[ECMB], mat+i, modestya, MOSISOLID.
Universidade de Vigo	GSC, INFERES, MAI.
Universitat de Girona	CODA.
Universitat de Lleida	CG, SSD.
Universitat Jaume I de Castellón	MODESMAN.
Universitat Politècnica de Catalunya	GNOM, GRASS, PROMALS, varidis.

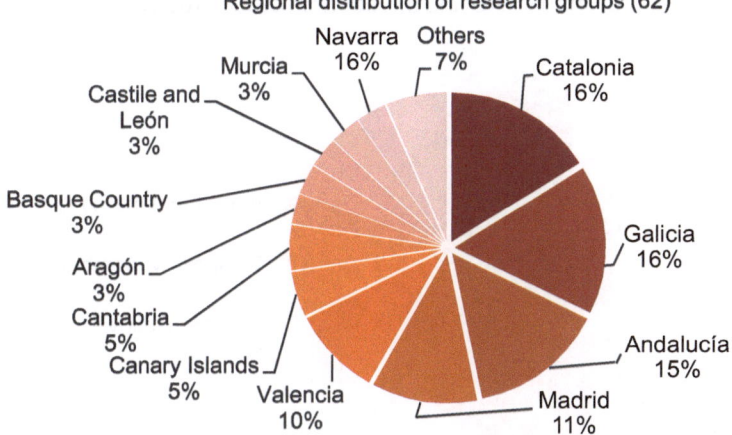

Fig. 3.2 Regional percentage distribution for all research groups in the Consulting Platform. The total number of research groups is indicated in brackets

3.1.2.3
Characterisation of research groups: staff distribution

The 62 research groups that took part in the survey bring together over 620 researchers and support staff. Regarding the distribution of members of staff (permanent and temporary staff, and research fellows), the staff percentage distribution is shown in Fig. 3.3.

The majority of staff are permanent members, with a figure of 472 (76%), compared to 73 temporary members of staff and 75 research fellows (12% each). In fact, 58% of research groups include research fellows and 44% of groups include temporary members of staff.

3.1.2.4
Characterisation of research groups: main MSC research areas

Figure 3.4 allows us a brief analysis of the main research areas where groups which belonged to the Consulting Platform have experience in technological supply to industry by means of competitive research projects, direct contracts with companies, or training.

It can be seen that the main research areas correspond to *Operations research and mathematical programming* (37%, 23 groups), *Statistics* (31%, 19 groups), and *Numerical analysis* (27%, 17 groups). In order to summarise, Fig. 3.4 only presents those MSC (Mathematics Subject Classification) research areas with more than 5 groups involved. Complete information and more details about MSC classification can be found in Sect. 3.3 of this chapter.

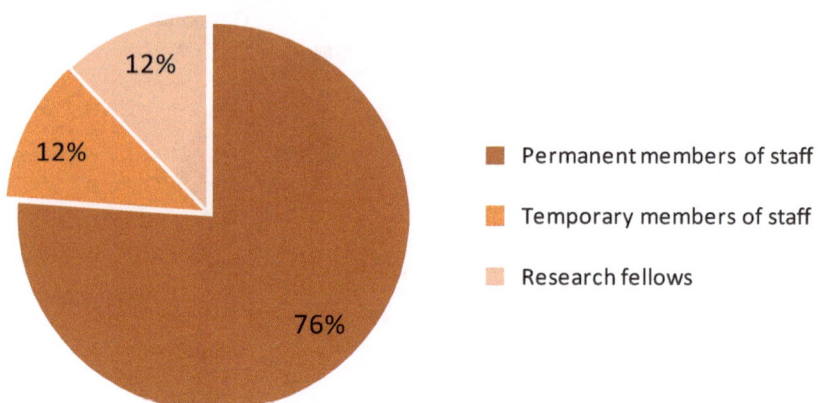

Staff distribution for the 62 research groups (620)

- Permanent members of staff
- Temporary members of staff
- Research fellows

Fig. 3.3 Staff percentage distribution for all research groups in the Consulting Platform. The total number of researchers and support staff is indicated in brackets

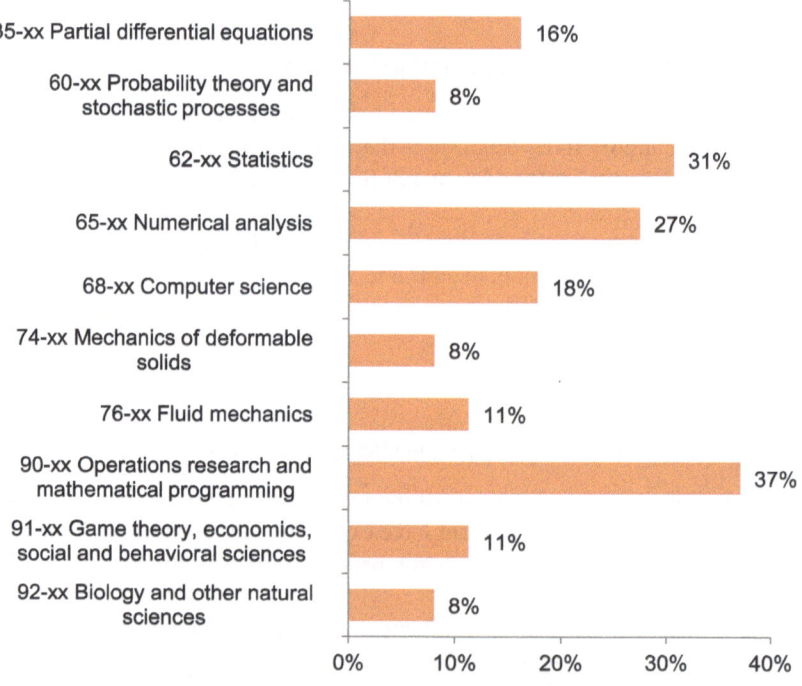

■ Percentages of research groups with experience by area (62)

Fig. 3.4 Percentages of research groups in the Consulting Platform with experience by MSC research area. The figure is restricted to MSC research areas where more than 5 research groups have stated they have experience. The total number of research groups in the Consulting Platform is indicated in brackets

3.2
Overview of research groups

Basic information about the 62 research groups with supply and proven experience in transfer of knowledge to the manufacturing and industrial sector, who took part in the survey for the TransMATH Supply Map, is given below.

The description of each group includes contact information (principal researcher, postal address, e-mail, and webpage), research lines, and the main keywords or key phrases describing their technological provision.

| AALN | Accurate and stable algorithms in Numerical Linear Algebra |
| | Algoritmos precisos y estables en Álgebra Lineal Numérica |

Froilán Martínez Dopico
Universidad Carlos III de Madrid

@ dopico@math.uc3m.es

⌨ gauss.uc3m.es

✉ Dpto. de Matemáticas
Campus de Leganés
Avda. Universidad 30
28911 Madrid

Research lines
- High precision algorithms for problems of eigenvalues, singular values, and matrix factorisation in structure based problems.
- Matrix perturbation theory with applications for stability analysis.

| ACEIA | Computational Algebra and Artificial Intelligence |
| | Álgebra Computacional e Inteligencia Artificial |

Eugenio Roanes Lozano
Universidad Complutense de Madrid

@ eroanes@mat.ucm.es

⌨ nash.sip.ucm.es/GrupoInvestACEIA

✉ Dpto. de Álgebra
Facultad de Educación
c/ Rector Royo Villanova s/n
28040 Madrid

Research lines
- Applications in logic and RBES.
- Applications in transportation engineering.
- Connection between Dynamic Geometry Systems (DGS) and symbolic computation systems (CAS).
- Intelligent behaviour.
- Architecture for changes in representation and Cognitive Science.
- Applications in automatic proof.
- Discovery of geometrical theorems.

Keywords/ Keyphrases Symbolic computation. Dynamic Geometry.

ADF — Fluid dynamics area / Área de dinámica de fluidos

Fernando Monge Gómez
Instituto Nacional de Técnica Aeroespacial

@ mongef@inta.es

⌨ www.inta.es

✉ Dpto. de Aerodinámica y Propulsión
Carretera de Ajalvir km. 4.5
Torrejón de Ardoz
28850 Madrid

Research lines
- Development of Computational Fluid Dynamics (CFD) codes.
- Numerical methods.
- Simulations of fluids.
- High performance computation.

Keywords/ Keyphrases   Computational fluid dynamics (CFD). Aerodynamic simulation. High performance computation. Renewable energy.

CAG — Computational Algebra and Geometry / Álgebra y Geometría Computacional

Tomás Recio Muñiz
Universidad de Cantabria

@ tomas.recio@unican.es

⌨ www.matesco.unican.es/grupos/Cag

✉ Dpto. de Matemáticas, Estadística
y Computación
Facultad de Ciencias
Avda. de los Castros s/n
39071 Santander

Research lines
- Development and analysis of efficacy, of symbolic/numerical algorithms, cryptography and numerical-symbolic scientific computation.
- Development of software and computer-assisted geometric design.

Keywords/ Keyphrases   Computer-assisted Geometric Design and CAD/CAM. Intersection problems of geometric entities. Network modelling. Resolution, with or without parameters, of nonlinear equations systems. Development of software and scientific calculus about symbolic-numerical algorithms.

CG — Cryptography and Graphs / Criptografía y Grafos

Josep M. Miret Biosca
Universitat de Lleida

@ miret@matematica.udl.cat

✉ c/ Jaume II 69
25001 Lleida

Research lines
- Cryptography.
- Graph theory and applications.
- Security and privacy in smart cards and RFID systems.
- Electronic voting.

Keywords/   Cryptography. Security. e-voting. Graphs.
Keyphrases

CODA	Statistics and Data Analysis Research Group
	Grupo de Investigación de Estadística y Análisis de Datos

José Antonio Martín Fernández
Universitat de Girona

@ josepantoni.martin@udg.edu   ⊠ Dpto. de Informática y Matemática
▦ ima.udg.es/Recerca/EIO/inici_esp.html   Aplicada
                                       Campus Montilivi
                                       Edificio P-IV
                                       17071 Girona

Research lines
- Statistical analysis of compositional data.

Keywords/   Compositional data. Multivariate statistical analysis.
Keyphrases

CYOPT	Control and Optimization
	Control y Optimización

Eduardo Casas Rentería
Universidad de Cantabria

@ eduardo.casas@unican.es   ⊠ ETS de Ingenieros Industriales y de Telecomu-
                                   nicación
                                   Avda. de los Castros s/n
                                   39005 Santander

Research lines
- Optimal control of systems governed by differential equations.
- Optimization algorithms.
- Inverse problems associated with differential equations.

Keywords/   Resource optimization. Production planning. Systems control. Ordinary differential equa-
Keyphrases   tions. Partial differential equations.

DATFUN	Probabilistic metrics and functional data
	Métricas probabilísticas y datos funcionales

@ Juan Antonio Cuesta Albertos   ⊠ Dpto. de Matemáticas
  Universidad de Cantabria   Facultad de Ciencias
                                         Avda. de los Castros s/n
▦ cuestaj@unican.es   39071 Santander

Research lines

- Functional data.
- Robust statistics.
- Random projections.

DDA	SIANI – Modelling and Computational Simulation SIANI – Modelización y Simulación Computacional

Rafael Montenegro Armas y Gustavo Montero García
Instituto Universitario de Sistemas Inteligentes y Aplicaciones Numéricas en Ingeniería (SIANI), Universidad de Las Palmas de Gran Canaria

@ rafa@dma.ulpgc.es

▦ dca.iusiani.ulpgc.es/proyecto2008–2011

✉ SIANI
Campus Universitario de Tafira
Edificio Central del Parque Científico-Tecnológico
35017 Las Palmas de Gran Canaria

Research lines

- Automatic generation of triangular and tetrahedral meshes.
- Resolution of large systems of sparse linear equations.
- Modelling and simulation of 3D wind fields.
- Atmospheric contamination.
- Development of a model for the simulation of forest fires.
- Solar radiation models.
- Environmental prediction models.

Keywords/ Keyphrases Mesh generation and adaptation. Resolution of equations systems. Parameter estimation. Genetic algorithms in parallel. Wind simulation. Wind maps. Wind energy potential. Solar radiation. Solar maps. Prediction models.

DECYL	Data, Statistics, Quality and Logistics Datos, Estadística, Calidad y Logística

Fermín Mallor Giménez
Universidad Pública de Navarra

@ mallor@unavarra.es

✉ Dpto. de Estadística e Investigación Operativa
Campus Arrosadía
31006 Pamplona

Research lines

- Optimization with simulation.
- Operations research applied to industrial processes.
- Operations research applied to health care.
- Operations research applied to energy systems.
- Optimal design of experiments.
- Reliability models.

Keywords/ Keyphrases Optimization. Simulation. Data mining. Energy systems. Reliability. Quality. Industrial management.

DEPREN	Sport Performance
	Deporte Rendimiento

Javier Sampedro Molinuevo
Universidad Politécnica de Madrid

@   javier.sampedro@upm.sp

⌨ www.inef.upm.es

✉ Facultad de las Ciencias de la Actividad
    Física y el Deporte
    Avda. Martín Fierro s/n
    28040 Madrid

Research lines
- Dynamic systems.
- Stochastic optimization.
- Analysis of temporal series.
- Applications for biomechanics and exercise physiology.
- Game analysis.
- Youth health and exercise.

Keywords/ Keyphrases	Dynamic systems. Stochastic optimization. Exercise physiology. Biomechanics. Game analysis.

EDANYA	Differential Equations, Numerical Analysis and Applications
	Ecuaciones Diferenciales, Análisis Numérico y Aplicaciones

Carlos Parés Madroñal
Universidad de Málaga

@   pares@anamat.cie.uma.es

⌨ edanya.uma.es

✉ Dpto. de Análisis Matemático
    Facultad de Ciencias
    Campus de Teatinos s/n
    29071 Málaga

Research lines
- Theoretical aspects: development of high-order numerical schemes for the approximation of weak solutions of hyperbolic systems with finite volume methods; treatment of source terms and non-conservative products, shock capturing, and well-balanced schemes.
- Applications: numerical schemes for shallow water models and related systems (multiphase or multilayer models, models including sediment and/or pollutant transport, underwater landslides, etc.). These models are applied to the simulation of geophysical fluids: rivers, lakes, or channel flows, marine currents, floods, tsunami generation and propagation, etc. (see http://anamat.cie.uma.es/animaciones).
- Vector and parallel computations: parallelization on PCs clusters. Exploitation of the multimedia records offered by current processors to perform a SIMD parallelization on each node of the cluster. Graphics cards (GPUs) programming. Development of the web platform HySEA that allows the user to simulate geophysical flows by using a standard browser as an interface between user and supercomputer.

Keywords/ Keyphrases	Urban hydraulics, water resources, marine hydraulics. Risk prevention. Oceanography. Finite element methods, finite volume methods, high order methods. Mesh generation and adaptation. Non-linear partial differential equations. Computational fluid dynamics, shallow water models. Vector and parallel programming, GPU programming, 3D visualization.

EDnL	Non Linear Differential Equations
	Ecuaciones Diferenciales no Lineales

Juan José Nieto Roig
Universidade de Santiago de Compostela

@ juanjose.nieto.roig@usc.es

⌨ www.usc.es/ednl

✉ Dpto. de Análisis Matemático
Facultad de Matemáticas
c/ Lope Gómez de Marzoa s/n
15782 Santiago de Compostela

Research lines
- Ordinary differential equations.
- Functional equations.
- Difference equations.
- Differential partial equations.
- Mathematical biology.
- Bioinformatics.

Keywords/ Keyphrases   Ordinary differential equations. Functional equations. Equations in differences. Partial differential equations. Mathematical biology. Bioinformatics. Symbolic computation. Green functions.

EE	Spatial Statistics
	Estadística Espacial

María Dolores Ugarte Martínez
Universidad Pública de Navarra

@ lola@unavarra.es

⌨ www.unavarra.es/personal/lugarte

✉ Dpto. de Estadística e I.O.
Campus de Arrosadía
Edificio de los Magnolios, primera planta
31006 Pamplona

Research lines
- Small area estimation.

Keywords/ Keyphrases   Small area estimation. Statistical modelling applied to various fields.

EOPT	Statistics and Optimization
	Estadística y Optimización

María Araceli Garín Martín
Universidad del País Vasco – Euskal Herriko Unibertsitatea

@ mariaaraceli.garin@ehu.es

⌨ www.et.bs.ehu.es/~eopt

✉ Dpto. de Economía Aplicada III
Facultad de Ciencias Económicas y
Empresariales
Avda. Lehendakari Aguirre
48015 Bilbao

Research lines
- Multivariate analysis and its applications, principally in questions of Marketing.
- Algorithmic research, particularly in 0–1 mixed optimization, both deterministic and under uncertainty.
- Applications in production planning, in supply chains and logistics (assignation of shift work in large organizations, optimization of storage and supply, optimization of energy transportation and distribution networks).
- Imputation of missing data and its applications for non-response in surveys, registry fusion, etc.
- Imputation in time series.
- Applications in relation to housing price indices, sales forecasts, or statistical operation revisions relating to Industrial Production Index (IPI) and surveys of R&D activity.

Keywords/ Multivariate analysis. Market segmentation. Time series. Missing data. Statistical imputa-
Keyphrases tion. Optimization applications. Mathematical programming. Index numbers.

Funaphy	Functional Analysis and Applications to Physics

Enrique A. Sánchez Pérez
Universidad Politécnica de Valencia

@ easancpe@mat.upv.es

⌨ www.upv.es/funaphy

✉ Instituto de Matemática Pura y Aplicada
Edificio IDI5 (8E), Cubo F, cuarta planta
46183 Valencia

Research lines
- Functional analysis.
- Banach lattices and Banach function spaces.
- Vector measurements.
- Mathematical methods in acoustics.
- Mathematical modelling in didactics.

Keywords/ Approximation of functions. Signal processing. Mathematical models in acoustics.
Keyphrases

GAUCA	UCA Algebra Group Grupo de Álgebra de la UCA

Enrique Pardo Espino
Universidad de Cádiz

@ enrique.pardo@uca.es

✉ Dpto. de Matemáticas
Apdo. 40
11510 Puerto Real

Research lines
- Stable and non-stable K theory on exchange rings.
- Leavitt path algebra structure on directed graphs.
- Monoidal structure and pre-ordered abelian groups satisfying Riesz's interpolation property.

GEUVA   UVA Statistical Applications Group
Grupo de Aplicaciones Estadísticas de la Universidad de Valladolid

Pedro César Álvarez Esteban
Universidad de Valladolid

@ pedroc@eio.uva.es

▦ www.eio.uva.es

✉ Dpto. de Estadística e Investigación Operativa
Facultad de Ciencias
Avda. Prado de la Magdalena s/n
47005 Valladolid

Research lines
- Reliability of complex systems.
- Risk evaluation.
- Biostatistics.
- Bioinformatics.
- Power efficiency.

- Prediction models.
- Data Mining.
- Quality control.
- Processes control.
- Sampling and resampling.
- Statistical methods in Demography.

GIO   Operations Research Group
Grupo de Investigación Operativa

Blas Pelegrín Pelegrín
Universidad de Murcia

@ pelegrin@um.es

▦ www.um.es/geloca/gio

✉ Dpto. de Estadística e Investigación Operativa
Facultad de Matemáticas
Campus de Espinardo
30100 Murcia

Research lines
- Localization and distribution models (discrete, continuous, in networks).
- Optimization techniques (global and discrete).
- Decision-making with various objectives.

Keywords/ Keyphrases Optimization of resources. Location of services. Planning. Transport and distribution. Logistics.

GIOPTIM   Optimization Research Group
Grupo de Investigación en Optimización

Rafael Blanquero Bravo
Universidad de Sevilla

@ rblanquero@us.es

▦ www.grupo.us.es/gioptim

✉ Dpto. de Estadística e Investigación Operativa
Facultad de Matemáticas
Campus de Reina Mercedes
41012 Sevilla

Research lines

● Mathematical and statistical modelling for decision-making.

Keywords/   Optimization. Decision-making. Data mining.
Keyphrases

| GIOS | Optimization and Simulation Research Group |
| | Grupo de Investigación Optimización y Simulación |

Herminia I. Calvete Fernández
Universidad de Zaragoza

@   herminia@unizar.es   ✉ Dpto. de Métodos Estadísticos
                                               Facultad de Ciencias
                                               Pedro Cerbuna 12
                                               50009 Zaragoza

Research lines

● Bilevel optimization.
● Multiobjective optimization.
● Vehicle routing problems.
● Logistics.
● Metaheuristics algorithms.
● Evolutive algorithms.
● Computational geometry and graphs theory.
● Optimizing and simulating processes: design and implementation of algorithms.

Keywords/   Optimization. Supply chain. Logistics. Simulation.
Keyphrases

| GMFN | Natural Phenomena Modelling Group |
| | Grupo de Modelado de Fenómenos Naturales |

Juan Luis Fernández Martínez
Universidad de Oviedo

@   jlfm@uniovi.es   ✉ Dpto. de Matemáticas.
                                               Facultad de Ciencias
                                               c/ Calvo Sotelo s/n
                                               33007 Oviedo

Research lines

● Geostatistical modelling.
● Inverse problems in environmental geophysics.
● Hydro-geological modelling.
● Modelling of geological structures.
● Application in earth sciences and applications for the characterisation and resolution of environmental problems.

Keywords/   Environmental models. Environmental geophysics. Hydro-geological models. Non-
Keyphrases   destructive inspection techniques. Geostatistical models.

GNOM   Numerical Optimization and Modelling Group

Jordi Castro Pérez
Universitat Politècnica de Catalunya

@ jordi.castro@upc.edu

⌨ gnom.upc.edu

✉ Dpto. de Estadística e Investigación Operativa
Campus Nord
Jordi Girona 1–3
Office C5 203
08034 Barcelona

Research lines
- Linear, nonlinear and integer optimization.
- Mathematical Programming.
- Applications in statistical disclosure control.
- Applications in energy.
- Applications in engineering.

GOMA   Applied Mathematics Optimization Group
Grupo de Optimización Matemática Aplicada

Juan José Salazar González
Universidad de La Laguna

@ jjsalaza@ull.es

⌨ www.goma.ull.es

✉ Dpto. de Estadística, Investigación Operativa y Computación
Facultad de Matemáticas
La Laguna
38271 Santa Cruz de Tenerife

Research lines
- Combinatory optimization.
- Heuristics.
- Vehicle routing.
- Telecommunications.
- Service location.
- Information protection.

Keywords/ Keyphrases   Limited resources optimization. Vehicle routing. Transport. Logistics. Heuristics.

GOR   Resources Optimization Group
Grupo de Optimización de Recursos

Alfredo Marín Pérez
Universidad de Murcia

@ amarin@um.es

⌨ www.um.es/or

✉ Facultad de Matemáticas
Campus de Espinardo
30100 Murcia

Research lines
- Combinatory optimization.
- Localization.
- Combinatory auctions.

GPB97	Statistics and Operations Research Modelling
	Modelos de Estadística e Investigación Operativa

Justo Puerto Albandoz
Universidad de Sevilla

@ puerto@us.es

⌨ www.grupo.us.es/fqm331

✉ Dpto. de Estadística e Investigación Operativa
Facultad de Matemáticas
c/ Tarfia s/n
41012 Sevilla

Research lines
- Localization.
- Analysis of distribution chains.
- Game theory.
- Multi-objective optimization.

Keywords/ Keyphrases Assistance in decision-making. Optimization of resources. Logistics and transport.

GRASS	Statistical Survival Analysis Research Group
	Grup de Recerca en Anàlisi Estadística de la Supervivència

Guadalupe Gómez Melis
Universitat Politècnica de Catalunya

@ lupe.gomez@upc.edu

⌨ grass.upc.edu

✉ Dpto. de Estadística e Investigación Operativa
Jordi Girona 1–3
08034 Barcelona

Research lines
- Biostatistics.
- Bioinformatics.
- Survival analysis.
- Building durability.

Keywords/ Keyphrases Regression models. Survival. Logistic regression.

GRID[ECMB]	Interdisciplinary Group in Statistics, Computing, Medicine and Biology
	Grupo Interdisciplinar de Estadística, Computación, Medicina y Biología

Carmen Cadarso Suárez
Universidade de Santiago de Compostela

@ gridecmb@gmail.com

⌨ eio.usc.es/pub/gridecmb

✉ Unidad de Bioestadística
Dpto. de Estadística e Investigación
Operativa
Facultad de Medicina
c/ San Francisco s/n
15782 Santiago de Compostela

Research lines

- Theoretical statistics (aimed at biomedical problems), computational statistics (techniques of computational acceleration and software development) and applied statistics research (interdisciplinary with specialists in Medicine and Biology) on GAM models and extensions of interest. Specifically, statistical inference is performed in generalised additive models and their derivates, these models being appropriately adjusted to association, prediction and classification studies.
- ROC regression techniques and quantile regression.
- Generalisations of these models: (i) GAMs including factor-by-curve interactions, curve-by-curve (surface) and factor-by-surface; (ii) unknown link functions; (iii) vectorial multi-dimensional response GAMs; (iv) multistate additive survival models.

Keywords/   Regression models. Factor-by-curve interactions. Firing rate. Temporal synchrony.
Keyphrases

GSC	Simulation and Control Group
	Grupo de Simulación y Control

Lino José Álvarez Vázquez
Universidade de Vigo

@ lino@dma.uvigo.es

⌨ www.dma.uvigo.es/~lino

✉ Dpto. de Matemática Aplicada II
ETSI Telecomunicaciones
Campus Lagoas-Marcosende
36310 Vigo

Research lines

- Optimization and optimal control.
- Mathematical modelling.
- Environmental engineering.
- Food engineering.
- Telecommunication systems.
- Asymptotic analysis.
- Mathematical elasticity.

Keywords/   Sterilisation. Pollution. Optimization. Simulation. Modelling. Control.
Keyphrases

| GSCUPM | Complex Systems Group |
| | Grupo de Sistemas Complejos |

Rosa María Benito Zafrilla
Universidad Politécnica de Madrid

@ rosamaria.benito@upm.es

www.gsc.upm.es

✉ Dpto. de Física y Mecánica
ETSI Agrónomos
Ciudad Universitaria
Avda. Complutense s/n
28040 Madrid

Keywords/ Signal analysis. Time series. Complex networks. Cellular automata.
Keyphrases

| GSD | UAB Group in Discrete Dynamical Systems |
| | Grupo de Sistemas Dinámicos Discretos de la UAB |

Lluís Alsedà i Soler
Universidad Autónoma de Barcelona

@ alseda@mat.uab.es

www.gsd.uab.es

✉ Dpto. de Matemáticas
Bellaterra
08193 Barcelona

| GSO | Optimization Solutions Group |
| | Grupo de Soluciones de Optimización |

Pedro César Álvarez Esteban
Universidad de Valladolid

@ pedroc@eio.uva.es

www.eio.uva.es

✉ Dpto. de Estadística e Investigación
Operativa
Facultad de Ciencias
Avda. Prado de la Magdalena s/n
47005 Valladolid

Research lines
- Location of services (health, distribution, etc.).
- Network design.
- Staff planning (timetables, shifts, etc.) and optimization of schedules.
- Production and transportation systems planning.
- Heuristics and customized algorithms.
- Applications for the health sector.

HYPCHAOP   Hypercyclicity and operators chaos
Hiperciclicidad y caos de operadores

Alfred Peris Manguillot
Universidad Politécnica de Valencia

@   aperis@mat.upv.es

✉   Dpto. de Matemática Aplicada
Instituto de Matemática Pura y Aplicada
Edificio 7A
46022 Valencia

Research lines
- Dynamical systems and chaos for operators.
- Signal and image processing.

Keywords/ Keyphrases   Diagnosis of failures in industrial engines. Mathematical models for processes of ablation through radio-frequency and laser. Sustainable land subdivision, optimal grouping of lands.

INFERES   Statistical Inference, Decision & Operations Research
Inferencia Estadística, Decisión e Investigación Operativa

Jacobo de Uña Álvarez
Universidade de Vigo

@   jacobo@uvigo.es

⌨   sidor.uvigo.es

✉   Dpto. de Estadística e Investigación Operativa
Campus Lagoas-Marcosende
36310 Vigo

Research lines
- Statistical modelling.
- Survival analysis.
- Time series.
- Spatial statistics.
- Nonparametric and semiparametric inference.
- Bioinformatics.
- Applications (Medicine, Economy, Engineering, Environment, Biology, Oceanography, etc.).
- Cost assignment.
- Route optimization.
- Models in Operations Research.
- Game theory.

Keywords/ Keyphrases   Data analysis. Survival analysis. Statistical assessment. Biostatistics. Econometry. Software development. Surveys and macrosurveys. Spatial statistics. Environmental statistics. Geostatistics. Statistical inference. Statistical modelling. Statistical packages. Statistical prediction. Decision-making. Route optimization. Cost sharing. Analysis of non-cooperative situations. Instructional design. e-learning.

InterTech	Interdisciplinary Modelling Group Grupo de Modelización Interdisciplinar

Pedro Fernández de Córdoba Castellá
Universidad Politécnica de Valencia

@ pfernandez@mat.upv.es

⌨ www.intertech.upv.es

✉ Dpto. de Matemática Aplicada
Camino de Vera s/n
46022 Valencia

Research lines
- Synthetic biology.
- Photonics.
- Energy engineering.
- Mathematical modelling.
- Numerical methods.

Keywords/ Keyphrases Advanced mathematical models for heat transfer (industrial reconditioning, geothermal heat pump). Synthetic biology and metabolic models. Electromagnetic behaviour models in micrometric and submicrometric scale photonic systems and devices.

KINETIC	Qualitative Properties of Partial Differential Equations in Kinetics and Diffusion Propiedades Cualitativas de Ecuaciones en Derivadas Parciales Cinéticas y de Difusión

José A. Carrillo de la Plata
Universidad Autónoma de Barcelona

@ carrillo@mat.uab.es

⌨ kinetic.mat.uab.es/research.html

✉ Dpto. de Matemáticas
Campus de la UAB
8193 Barcelona

Research lines
- Numerical methods for kinetic equations and nonlinear diffusion equations.

Keywords/ Keyphrases Large particle systems. Electronic Engineering.

LOGRO	Localization Research Group Grupo de Investigación en Localización

Juan Antonio Mesa López-Colmenar
Universidad de Sevilla

@ jmesa@us.es

⌨ grupo.us.es/logro

✉ Dpto. de Matemática Aplicada II
Escuela Técnica Superior de Ingenieros
Camino de los Descubrimientos s/n
(Enríquez de Ribera 1)
41092 Sevilla

Research lines
- Facility location.
- Routes.
- Network design.
- Transport models.
- Data classification.
- Logistics.

Keywords/ Keyphrases   Transport optimization. Localisation. Network design.

M2NICA	Numerical Models and Methods in Engineering and Applied Sciences
	Modelos y Métodos Numéricos en Ingeniería y Ciencias Aplicadas

Carlos Vázquez Cendón
Universidade da Coruña

@ carlosv@udc.es

📖 dm.udc.es/m2nica

✉ Dpto. de Matemáticas
Facultad de Informática
Campus Elviña s/n
15071 A Coruña

Research lines
- Modelling.
- Mathematical analysis.
- Numerical simulation of problems in Engineering and Applied Sciences.

Keywords/ Keyphrases   Mathematical models. Solid and fluid mechanics. Acoustics. Quantitative finance. Partial differential equations. Numerical methods. Finite elements. Numerical simulation. Computer applications. High performance computing. GPU's. Free software for science and engineering.

M2S2M	Mathematical Modelling and Simulation of Environmental Systems
	Modelado Matemático y Simulación de Sistemas Medioambientales

Tomás Chacón Rebollo
Universidad de Sevilla

@ chacon@us.es

📖 grupo.us.es/gfqm120/web_grupo

✉ Dpto. de Ecuaciones Diferenciales y
Análisis Numérico
Facultad de Matemáticas
c/ Tarfia s/n
41012 Sevilla

Research lines
- Numerical modelling of shallow water flows.
- Numerical modelling of oceanic flows.
- Domain decomposition techniques.
- Tsunami modelling and simulation.
- Environmental risk assessment.

Keywords/ Keyphrases   Hydrodynamic flows. Numerical simulation. Flooding. Risk prevention. Port engineering.

MAI	Differential Equations and Numerical Simulation Group
	Grupo de Ecuaciones Diferenciales y Simulación Numérica

José Durany Castrillo
Universidade de Vigo

@ durany@dma.uvigo.es

⌨ www.dma.uvigo.es

✉ Dpto. de Matemática Aplicada II
ETSI Telecomunicaciones
Campus Lagoas-Marcosende
36310 Vigo

Research lines
- Numerical simulation of physical and engineering processes governed by differential equations and partial differential equations.
- Boundary problems for nonlinear differential equations.
- Differential equations with impulses.
- Integer-differential equations.

Keywords/ Keyphrases   Engineering. Equations. Physics. Mathematics. Automotive. Modelling. Simulation.

mat+i	Research Group in Mathematical Engineering
	Grupo de Investigación en Ingeniería Matemática

Alfredo Bermúdez de Castro López-Varela
Universidade de Santiago de Compostela

@ alfredo.bermudez@usc.es

⌨ www.usc.es/ingmat

✉ Dpto. de Matemática Aplicada
Facultad de Matemáticas
Campus Vida
15782 Santiago de Compostela

Research lines
- Modelling and numerical simulation in Engineering, Finance and Environment: solid mechanics (structural engineering, large deformations, friction, contact, fracture), fluid mechanics (hydraulics, gas dynamics), electromagnetism (eddy current, induction furnaces, electrical machinery, non destructive testing), acoustics (sound level, sound insulation, passive and active control of noise, sound absorbing materials), heat transfer (conduction, convection and radiation), chemical reactions (equilibrium, finite kinetics), combustion (power stations boilers), fluid-structure interaction (design of sails and kites), mechanical and acoustic-structural vibrations, evaluation of financial derivatives.
- Numerical simulation of coupled problems (multiphysics).
- Homogenization of periodic materials.
- Optimization and control of devices and processes.
- Inverse problems.
- Analysis of linear and nonlinear partial differential equations (PDE).
- Numerical resolution of PDE: finite element methods (FEM and XFEM), boundary integral methods (BEM) and reduced-order methods.

Keywords/ Keyphrases   Metallurgy: numerical simulation of metal and ferro-alloy casting, metallurgic electrodes, aluminium electrolysis in reduction pots, induction ovens and material purification. Solid mechanics: simulation of thin structures, solids, with contact, cracks and fracture, viscoelastic and viscoplastic behaviour laws. Energy: simulation of combustion in carbon or fuel boilers and problems in plasma physics. Environment: hydrodynamic flow simulation in aqueous media, dispersion of pollutants and chemical kinetics. Acoustic: simulation of fluid structure interaction problems, insulation materials, vibration and active noise control. Fluid mechanics: simulation of sails, heat transport by convection, diffusion transport at high temperatures and study of St. Venant equations. Automotive: simulation of burns by airbags, bus rollover simulation, noise control. Finance: valuation of financial products.

MCS-UAB	Mathematical Consulting Service
	Servei de Consultoria Matemàtica

Aureli Alabert Romero
Universidad Autónoma de Barcelona

@   aureli.alabert@uab.cat
      director@mcs-uab.com

⌨   www.mcs-uab.com

✉ Dpto. de Matemáticas
   Facultad de Ciencias
   Edificio C
   08193 Bellaterra

Research lines
- Stochastic optimization.
- Social choice.
- Stochastic differential equations.
- Classification.
- Multivariate modelling.

Keywords/   Modelling. Computation. Simulation.
Keyphrases

MODES	Modelling and Statistical Inference
	Modelización e Inferencia Estadística

Ricardo Cao Abad y Juan Manuel Vilar Fernández
Universidade da Coruña

@   rcao@udc.es, eijvilar@udc.es

⌨   dm.udc.es/modes

✉ Dpto. de Matemáticas
   Facultad de Informática
   Campus de Elviña s/n
   15071 A Coruña

Research lines
- Non-parametric inference.
- Dependent data.
- Survival analysis.
- Application of Statistics to other sciences.
- Game theory.
- Operational research.

Keywords/   Data analysis. Survival and reliability analysis. Multivariate analysis. Experimental analysis
Keyphrases and design. Quality control. Spatial statistics. Non-parametric and free distribution methods. Actuarial loss models. Risk modelling. Time series. Statistical association techniques. Imputation techniques. Statistical inference techniques. Prediction techniques. Resampling techniques.

MODESI	Stochastic Models
	Modelos Estocásticos

Gerardo Sanz Sáiz
Universidad de Zaragoza

@ gerardo@unizar.es

✉ Dpto. de Métodos Estadísticos
Facultad de Ciencias
c/ Pedro Cerbuna 12
Edificio B
50009 Zaragoza

Research lines
- Probability.
- Applied probability.
- Stochastic processes.
- Probabilistic networks.
- Neuronal networks.
- Climate data analysis.
- Extreme data analysis.
- Modelling in Health Sciences.

Keywords/ Keyphrases
Data analysis. Design, development and analysis of surveys. Database exploitation. Data mining. Statistical models for medicine and environment. Market studies. Quality control. Reliability.

MODESMAN	Statistical Modelling of Environmental Problems Group
	Grupo de Modelización Estadística para Problemas Medioambientales

Jorge Mateu Mahiques
Universitat Jaume I de Castellón

@ mateu@mat.uji.es

⌨ www3.uji.es/~mateu

✉ Dpto. de Matemáticas
Escuela Superior de Tecnología y Ciencias
Experimentales
Campus Riu Sec
12071 Castellón

Research lines
- Multivariate statistics.
- Spatial statistics.
- Modelling by accurate spatial processes.
- Geostatistical modelling.
- Modelling of data in lattices.
- Modelling of phenomena that vary in space and/or time.
- Applications of modelling in atmospheric contamination and soil problems.
- New theoretical space-time covariance structures.

Keywords/ Keyphrases
Modelling of phenomena with space-time evolution. Modelling of fires. Modelling of aquifers. Neurophysiologic modelling (evoked potentials). Modelling of mortality charts. Stochastic processes.

modestya	Optimization Modelling, Decision, Statistics and Applications
	Modelos de Optimización, Decisión, Estadística y Aplicaciones

Wenceslao González Manteiga
Universidade de Santiago de Compostela

@ wenceslao.gonzalez@usc.es

▦ eio.usc.es/pub/gi1914

✉ Dpto. de Estadística e Investigación
Operativa
Facultad de Matemáticas
Campus Vida
15782 Santiago de Compostela

Research lines

- Prediction models.
- Categorical data.
- Sampling techniques.
- Biostatistics.
- Resampling techniques.
- Time series.
- Multivariate analysis.
- Regression models.
- Non-parametric inference.
- Censored data.
- Geostatistics.
- Applications for game theory.
- Voting and power indices.
- Assignation of costs and tariff rate design.
- Interactive OR models.
- Multi-criteria and multi-level decision making and programming.
- Mathematical programming.

Keywords/ Keyphrases   Statistical inference. Biostatistics. Decision-making models. Prediction models. Game theory. Operational research models. Geostatistics. Time series. Data analysis. Optimization of resources. Quality control. Control of processes. Sampling and resampling.

modsol	Solubility, Integrability and Chaos in Classical and Quantum Systems
	Solubilidad, Integrabilidad y Caos en Sistemas Clásicos y Cuánticos

Artemio González López
Universidad Complutense de Madrid

@ artemio@fis.ucm.es

✉ Dpto. de Física Teórica II
Avda. Complutense s/n
28040 Madrid

Research lines

- Soluble models in Quantum Physics.
- Spin chains.
- Integrable systems.

MOSISOLID	Mathematical Modelling and Numerical Simulation in Solid Mechanics
	Modelos Matemáticos y Simulación Numérica en Mecánica de Sólidos

Juan Manuel Viaño Rey
Universidade de Santiago de Compostela

@ juan.viano@usc.es

▦ www.usc.es/dmafm/grupo_viano

✉ Dpto. de Matemática Aplicada
Facultad de Matemáticas
c/ Lope Gómez de Marzoa s/n.
15782 Santiago de Compostela

Research lines
- Thin structures: modelling and calculation of structures composed of elastic, viscoelastic or viscoplastic beams, plates and shells.
- Mechanics of contact: mathematical modelling and numerical simulation of contact problems, friction, adhesion and wear in elastic, viscoelastic and viscoplastic structures.
- Biomechanics: numerical simulation and mathematical models of human jaw and bone formation.
- Mechanical design: car steering wheels, orthodontics.

Keywords/ Keyphrases	Differential equations. Numerical simulation. Computer-aided design. Computational mechanics. Elasticity, plasticity, viscoelasticity, viscoplasticity, piezoelectricity. Contact mechanics, friction, adhesion, wear. Beams, plates, sheets, vibrations. Finite elements. Asymptotic methods. Meshing. Biomechanics, biomathematics, orthodontics, bone formation.

OEDgroup	**Optimum Experimental Design Group**

Jesús López Fidalgo
Universidad de Castilla-La Mancha, Universidad de Almería, Universidad de Salamanca

@ jesus.lopezfidalgo@uclm.es

⌨ areaestadistica.uclm.es/oed

✉ Dpto. de Matemáticas
Escuela Superior de Ingeniería Industrial
Avda. Camilo José Cela 3
13071 Ciudad Real

Research lines
- Nonlinear models.
- Pharmacokinetic models.
- Multifactorial models.
- Compartmental models.
- Survival analysis and censored variables.

Keywords/ Keyphrases	Experimental design. Michaelis Menten models. Compartmental models. Arrhenius model.

Orel	**Resources Optimization, Statistics, Transport and Logistics** **Optimización de Recursos, Estadística, Transporte y Logística**

Antonio Manuel Rodríguez Chía
Universidad de Cádiz

@ antonio.rodriguezchia@uca.es

⌨ www.uca.es/grupos-inv/FQM355

✉ Dpto. de Estadística e Investigación Operativa
Facultad de Ciencias
Polígono Río San Pedro
11510 Puerto Real

Research lines
- Location of service centres.
- Distribution chains.
- Routing, transport and logistics.
- Combinatory optimization.
- Resource optimization and production chains.
- Management of distribution and logistical systems.
- Risk analysis.
- Distribution approximation.

Keywords/ Localisation. Distribution and logistics networks. Routing problems. Optimization of re-
Keyphrases sources.

PROMALS	Research Group in Mathematical Programming, Logistics and Simulation Grupo de Investigación en Programación Matemática, Logística y Simulación

Elena Fernández Aréizaga
Universitat Politècnica de Catalunya

@   e.fernandez@upc.edu   ✉ Dpto. de Estadística e Investigación Operativa
                                        Campus Nord
                                        Jordi Girona 1–3
                                        Office C5 208
                                        08034 Barcelona

Research lines
- Integer programming and combinatory optimization.
- Traffic simulation and management.
- Simulation of discrete events.

Keywords/ Transport. Logistics. Traffic. Discrete location.
Keyphrases

Psycotrip	Research Group in Programming and Symbolic Computation Grupo de Investigación de Programación y Cálculo Simbólico

Julio Rubio García
Universidad de La Rioja

@   julio.rubio@unirioja.es   ✉ Dpto. de Matemáticas y Computación
                                        c/ Luis de Ulloa s/n
⌨ esus.unirioja.es/psycotrip   26004 Logroño

Research lines
- Symbolic computation systems as software systems.
- Formal methods in software Engineering.

Keywords/ Computer Science consultancy.
Keyphrases

RiTO	Risk, Time & Optimization

Laureano F. Escudero Bueno
Universidad Rey Juan Carlos

@ laureano.escudero@urjc.es

✉ Dpto. de Estadística
c/ Tulipán s/n
28933 Madrid

Research lines

- Electricity: short-term generation planning ("unit commitment"); open market bidding optimization; capacity expansion planning.
- Pipelines: capacity expansion planning; exploitation planning; supply and distribution of petrochemicals (hydrocarbons, petroleum and chemical products); transport and distribution logistics; supply chain management; strategic, tactical and operational planning of production; revenue management.
- Public Administration and large industrial groups: resource allocation; investment and large industrial group selection.
- Finance: treasury management; immunisation of portfolios of fixed income securities.
- Risk analysis in: finance; supply chains; environment and agriculture; national security against terrorist attacks and natural disasters; air operations; watershed management; water distribution; emotional robots; electronic participation, cyber security and fraud detection.
- Planning: forestry harvesting; detection and resolution of air traffic conflicts; mineral extraction; planning, location and design of correctional facilities.
- Applications on real world problems nationally and internationally, mainly Europe, USA, Canada, Africa, and Latin America.

Keywords/ Keyphrases	Algorithms. Computational implementations. Multi-criteria optimization. Time series. Stochastic processes. Data mining. Clinical studies. Genetic analysis and pattern recognition. Optimal planning and management. Decision analysis. Negotiation analysis. Adversarial risk analysis. Discrete event simulation. Decision making in groups. Scenario analysis. Analysis and management of risk.

RUTYLO	Routing Problems and Location Algorithms Algoritmos para Problemas de Rutas y Localización

Enrique Benavent López
Universidad de Valencia

@ Enrique.Benavent@uv.es

✉ Dpto. de Estadística e Investigación Operativa
c/ Dr. Moliner 50
Burjasot
46100 Valencia

Research lines

- Routing problems.
- Localisation.

Keywords/ Keyphrases	Routing problems. Optimization in transport. Metaheuristics.

RUTYMETA	Routes and Metaheuristics
	Rutas y Metaheurísticos

Ángel Corberán Salvador
Universidad de Valencia

@ angel.corberan@uv.es   ✉ Dpto. de Estadística e Investigación Operativa
c/ Dr. Moliner 50
Burjasot
46100 Valencia

Research lines
- Routing problems.
- Metheuristics.

Keywords/   Routing problems. Optimization in transport. Metaheuristics.
Keyphrases

SSD	Dynamic Systems Seminar
	Seminari de Sistemes Dinàmics

Jaume Giné Mesa
Universitat de Lleida

@ gine@matematica.udl.cat   ✉ Dpto. de Matemática
⌨ www.ssd.udl.cat   Escuela Politécnica Superior
Avda. Jaume II 69
25001 Lleida

Research lines
- Ordinary differential equations.
- Qualitative theory of differential equations.

Keywords/   Symbolic manipulators. Programming languages.
Keyphrases

TAMI	Processing and Mathematical Analysis of Digital Images
	Tratamiento y Análisis Matemático de Imágenes Digitales

Bartomeu Coll Vicens
Universidad de las Islas Baleares

@ tomeu.coll@uib.es   ✉ Dpto. de Matemáticas e Informática
Carretera Valldemossa km. 7.5
7122 Palma de Mallorca

Research lines
- Processing, restoration and analysis of digital images based on mathematical models (PDEs, variational, statistical, etc.) with applications in satellite imaging and digital photography.

Keywords/   Processing, restoration and analysis of digital images and applications.
Keyphrases

TAPO	Approximation Theory and Orthogonal Polynomials Teoría de Aproximación y Polinomios Ortogonales

Andrei Martínez Finkelshtein
Universidad de Almería

@ andrei@ual.es

▦ www.ual.es/GruposInv/Tapo

✉ Dpto. de Estadística y Matemática
Aplicada
Facultad de Ciencias Experimentales
Carretera de Sacramento s/n
04120 Almería

Research lines
- Electrostatic interpretation and density of zeros of special functions families.
- Modern asymptotic analysis methods of special functions.
- Applications in random matrices and stochastic processes.
- Algebraic and asymptotic study of families of orthogonal polynomials regarding non-standard scalar products (in particular, Sobolev's scalar product).
- Study, applications and computation of information measures of special functions.
- Study of orthogonal polynomials of several variables.
- Methods for modelling and diagnostics in ophthalmology, including reconstruction techniques for the surface of human cornea from topographical data and early detection mechanisms for related pathologies.

Keywords/
Keyphrases
Zernike Polynomials. Surfaces reconstruction. Surface modelling. Corneal irregularities. Gaussian Functions. Radial base functions. Multi-scale methods.

TD-ULPGC	Bayesian and Decision Statistical Techniques in Economy and Enterprise Técnicas Estadísticas Bayesianas y de Decisión en Economía y Empresa

Dolores R. Santos Peñate
Universidad de Las Palmas de Gran Canaria

@ drsantos@dmc.ulpgc.es

▦ www.gi.ulpgc.es/tebadm

✉ Dpto. de Métodos Cuantitativos
Facultad de Ciencias Económicas y
Empresariales
Campus de Tafira
Módulo D, planta 4, despacho 22
35017 Las Palmas de Gran Canaria

Research lines
- Optimization.
- Localisation.
- Input-Output analysis.
- General equilibrium models.
- Spatial analysis and geographical information systems.
- Bibliometric analysis.

Keywords/
Keyphrases
Location and logistics. Accountancy matrices. Spatial analysis. Economic geography. Geographic information systems. General equilibrium. Bibliometric analysis.

TOREFA	Operator theory: Lattices and Analytical Function Spaces Teoría de Operadores: sus Retículos y Espacios de Funciones Analíticas

Alfonso Montes Rodríguez
Universidad de Sevilla

@ amontes@us.es

✉ Dpto. de Análisis Matemático
Apdo. 1160
41080 Sevilla

Research lines
- Operator theory.
- Analytic functions.
- Volterra-type operators.
- Composition operators.

Keywords/ Linear operator. Spectrum of an operator. Functions iteration.
Keyphrases

TTM	Mathematical Technology Transfer Group Grupo de Transferencia de Tecnología Matemática

Mikel Lezaun Iturralde
Universidad del País Vasco – Euskal Herriko Unibertsitatea

@ mikel.lezaun@ehu.es

▦ www.ehu.es/mae/grupottm

✉ Dpto. de Matemática Aplicada y
Estadística e Investigación Operativa
Facultad de Ciencia y Tecnología
Barrio Sarriena s/n
Leioa
48940 Vizcaya

Research lines
- Optimization.
- Logistics.
- Data processing.
- Numerical simulation.

Keywords/ Optimization. Data Processing.
Keyphrases

varidis	Discrete Manifolds and Potential Theory

Enrique Bendito Pérez
Universitat Politècnica de Catalunya

@   enrique.bendito@upc.edu   ⊠ Dpto. de Matemática Aplicada
                                                    Jordi Girona Salgado 1–3
                                                    08034 Barcelona

Research lines
- Approximation in potential theory.
- Discrete vector calculus.
- Numerical methods.

Keywords/   Non-linear optimization with constraints. Design of earthing networks. Information diffu-
Keyphrases   sion algorithms. Structure instrumentation. Generation of finite element meshes.

3.3
Mathematical techniques and MSC classification

The 62 groups involved in the survey are working in various areas of mathematics. Together they offer a large range of mathematical, operations research and statistical techniques, representing the experience they have gained over their many years of work.

The main mathematical techniques in which the groups demonstrate transfer experience are detailed below. All these techniques have been incorporated into the overall provision available for transferring mathematical techniques to industry, and have been classified in three main categories, which were considered in the Trans-MATH Demand Map (see Chap. 2):

- CAD/CAE: computer-aided design and numerical simulation, usually known as computer-aided engineering (see Table 3.2);
- ST/OR: statistical and operations research tools, such as data analysis techniques or decision-making support techniques (see Table 3.3);
- OMT: other mathematical techniques, for example, geographical tracking systems, signal processing, computation, biomathematics, etc. (see Table 3.4).

It should be noted that the following tables are not exhaustive lists of all the mathematical techniques that research groups understand and handle, but a brief overview of the main mathematical tools which they have used in direct contracts with industry and training courses in recent years.

Table 3.2 Research groups in the Consulting Platform with relevant transfer experience in CAD/CAE techniques

Category	Subcategory Groups
General	Mathematical modelling. EDnL, Funaphy, GMFN, GOMA, GSC, HYPCHAOP, InterTech, M2NICA, M2S2M, MAI, mat+i, MCS-UAB, MOSISOLID, OEDgroup, TAMI, TAPO, TTM.
	Ordinary differential equations and systems. CYOPT, EDnL, GSD, M2NICA, MAI, MCS-UAB, MOSISOLID, SSD.
	Partial differential equations: linear and non-linear analysis. Green's functions. ADF, CYOPT, EDANYA, EDnL, EOPT, GSC, HYPCHAOP, KINETIC, M2NICA, M2S2M, MAI, mat+i, MOSISOLID, TAMI, TTM.
	Operator theory. HYPCHAOP, TOREFA.
	Asymptotic methods. Homogenization. GSC, mat+i, MOSISOLID, TAPO.
	Optimal control. Optimization. AALN, ADF, CYOPT, Funaphy, GSC, mat+i, MOSISOLID, RiTO.
	Inverse problems. CYOPT, GMFN, mat+i.
	Function approximation techniques. Funaphy, MOSISOLID, TAPO.
	Difference and functional equations. EDnL, EOPT.
	Numerical methods. ADF, CAG, CYOPT, GSC, GSD, InterTech, KINETIC, M2NICA, MCS-UAB, MOSISOLID, RiTO, varidis.
	Finite difference methods. GMFN, mat+i, MOSISOLID.
	Finite element methods (FEM). DDA, EDANYA, Funaphy, GMFN, M2NICA, M2S2M, MAI, mat+i, MOSISOLID.
	Extended finite element methods (XFEM). mat+i.

continued…

Category	Subcategory Groups
	Finite volume methods. EDANYA, M2S2M, MAI, mat+i.
	Boundary element method (BEM). M2NICA, MAI, mat+i.
	Reduced-order methods. mat+i.
	Generation and adaptation of meshes. DDA, EDANYA, MOSISOLID, varidis.
	Numerical simulation. AALN, DECYL, GIOS, GMFN, GSC, GSD, M2NICA, M2S2M, MAI, mat+i, MCS-UAB, MODES, MOSISOLID, TTM.
Mechanical or structural	Structural engineering. Calculation of thin structures (beams, plates, sheets). Solids with contact, friction, adhesion models, wear, cracks and fractures. Elastic, viscoelastic and viscoplastic laws. Piezoelectric materials. Homogenization of periodic materials. Vibrations. Large deformations. GSC, GMFN, M2NICA, MAI, mat+i, MODESI, MOSISOLID.
Thermal or thermodynamics	Heat transfer: conduction, heat transport by convection, radiation. Diffusion transport at high temperatures. Study of St. Venant equations. DDA, GSC, HYPCHAOP, InterTech, M2NICA, MAI, mat+i, MOSISOLID.
Electronics and/or electromagnetics	Electronics. KINETIC, varidis
	Electromagnetism. Eddy current. InterTech, mat+i.
Fluids	Computational fluid dynamics (CFD). Hydraulic. Gas dynamic. Aerodynamic simulation. Shallow water models. Hydrodynamic flow simulation in aqueous media. Simulation of sails. Dispersion of pollutants. ADF, DDA, EDANYA, EOPT, GMFN, GSC, KINETIC, M2NICA, M2S2M, MAI, mat+i.
Acoustics or vibro-acoustics	Sound level. Sound insulation. Simulation of fluid-structure interaction. Sound absorbing materials. Vibration. Active and passive noise control. Funaphy, M2NICA, mat+i.
Combustion	Combustion processes. Power station boilers. DDA, MAI, mat+i.

continued...

Category	Subcategory Groups
Chemical kinetics	Finite kinetics. Equilibrium. KINETIC, mat+i, OEDgroup.
Risk & financial analysis	Quantitative finances. Risk modelling. Evaluation of financial derivatives. GSD, M2NICA, mat+i.
Biomechanics	Numerical simulation of human jaw and bone formation. Orthodontic models. MOSISOLID.
Geophysics	Geothermics. Oceanography. M2NICA, M2S2M, MAI.
Multiphysics	Fluid-structure interaction problems. Acoustic-structural vibrations. Thermo-mechanics. Thermo-hydrodynamics. Magneto-hydrodynamics. Magneto-thermics. GSC, M2NICA, MAI, mat+i.

Table 3.3 Research groups in the Consulting Platform with relevant transfer experience in ST/OR techniques

Category	Subcategory Groups
General	Statistical modelling. EE, GEUVA, GIOPTIM, GMFN, GPB97, INFERES, MCS-UAB, MODES, MODESI, MODESMAN, modestya, OEDgroup, Orel, RiTO, TAMI. Regression models. GEUVA, GRASS, GRID[ECMB], INFERES, MODES, MODESI, modestya. Prediction models. DDA, EE, GEUVA, GRID[ECMB], INFERES, MODES, MODESI, modestya, RiTO, TTM.
Quality control	Quality control. Reliability. DECYL, GEUVA, MODES, MODESI, MODESMAN, modestya.
Control & optimization of production, processes & stocks	Control of products. Control of processes. GEUVA, modestya, RiTO. Production planning. CYOPT, DECYL, EOPT, GIOS, modestya, PROMALS, RiTO, RUTYLO.

continued . . .

Category	Subcategory Groups
	Resource optimization. CYOPT, DECYL, EOPT, GIO, GIOS, GNOM, GOMA, GOR, GPB97, MCS-UAB, modestya, Orel, PROMALS, RiTO, RUTYLO.
Risk & financial analysis	Quantitative finances. Econometrics. Risk modelling. Evaluation of financial derivatives. EOPT, GEUVA, INFERES, MCS-UAB, MODES, modestya, Orel, RiTO, TD-ULPGC. General equilibrium. Social accounting matrices. EOPT, TD-ULPGC.
Strategy, logistics & planning	Decision-making problems. Decision theory. ACEIA, GIO, GIOPTIM, GIOS, GOR, GPB97, INFERES, MODES, modestya, PROMALS, RiTO. Logistics. Transport. ACEIA, CAG, DECYL, EOPT, GEUVA, GIO, GIOS, GOMA, GOR, GPB97, GSO, INFERES, LOGRO, MCS-UAB, modestya, Orel, PROMALS, RiTO, RUTYLO, RUTYMETA, TD-ULPGC, TTM. Routes planning and optimization. ACEIA, EOPT, GIO, GIOS, GOMA, GOR, GPB97, GSO, INFERES, LOGRO, MCS-UAB, modestya, Orel, PROMALS, RiTO, RUTYLO, RUTYMETA, TTM. Network modelling, planning and design. CAG, EOPT, GEUVA, GIO, GNOM, GSO, LOGRO, modestya, Orel, PROMALS, RiTO, RUTYLO, TTM, varidis. Location. GIO, GOMA, GOR, GPB97, GSO, LOGRO, Orel, PROMALS, RiTO, RUTYLO, TD-ULPGC.
Customer, market & product studies	Design, development and analysis of surveys. EE, EOPT, GOMA, INFERES, MCS-UAB, MODES, MODESI, modestya. Statistical confidentiality. Disclosure protection. GNOM, GOMA, GOR.
Exploitation of internal information	Data analysis. Data mining. DECYL, DEPREN, EE, EOPT, GEUVA, GIOPTIM, GIOS, INFERES, mat+i, MODES, MODESI, MODESMAN, modestya, OEDgroup, RiTO, RUTYMETA, TD-ULPGC, TTM.

continued...

Category	Subcategory Groups
Inference from stochastic processes	**Stochastic processes.** DATFUN, DECYL, EOPT, MODESI, MODESMAN, RiTO, TAPO. **Time series.** DEPREN, EOPT, GEUVA, GMFN, GSCUPM, INFERES, MCS-UAB, MODES, MODESI, MODESMAN, modestya, RiTO. **Spatial statistics.** CODA, GEUVA, GMFN, INFERES, MCS-UAB, MODES, MODESI, MODESMAN, modestya, TD-ULPGC. **Spatio-temporal statistics.** EE, GMFN, INFERES, MCS-UAB, MODES, MODESI, MODESMAN, modestya. **Neural nets.** Funaphy, HYPCHAOP, MODESI, modestya. **Functional data analysis.** DATFUN, modestya.
Parametric & nonparametric inference	**Statistical inference.** GRID[ECMB], INFERES, MODES, MODESMAN, modestya. **Parameter estimation.** DDA, INFERES, modestya, OEDgroup. **Small area estimation.** EE, GEUVA, MODES, modestya. **Robustness procedures.** DATFUN, EE, GSCUPM, LOGRO. **Resampling methods.** DATFUN, GEUVA, MODES, modestya, RiTO.
Multivariate analysis	**Multivariate analysis.** CODA, DATFUN, EE, EOPT, GEUVA, GMFN, INFERES, MCS-UAB, MODES, MODESI, MODESMAN, modestya, OEDgroup, TTM. **Classification and discrimination. Cluster analysis.** EOPT, GEUVA, GIOPTIM, GRID[ECMB], LOGRO, INFERES, MCS-UAB, MODES, MODESI, MODESMAN, modestya. **Compositional data analysis.** CODA.

continued...

Category	Subcategory Groups
Other statistical methods	Design of experiments. DECYL, MCS-UAB, MODES, OEDgroup, RUTYMETA.
	Sampling techniques. Imputation techniques. EE, EOPT, GEUVA, GIOPTIM, GRASS, INFERES, MODES, modestya.
	Survival analysis. EE, GRASS, GRID[ECMB], INFERES, MODES, MODESMAN, modestya, OEDgroup.
Operations research, mathematical programming & game theory	Operations research modelling and techniques. DECYL, GIO, GNOM, GOR, GPB97, INFERES, MCS-UAB, MODES, modestya, TD-ULPGC, TTM.
	Mathematical programming. Optimization. CAG, CYOPT, DECYL, DEPREN, EOPT, GIO, GIOPTIM, GIOS, GNOM, GOMA, GOR, GPB97, GSO, HYPCHAOP, INFERES, LOGRO, mat+i, MCS-UAB, MODESI, modestya, Orel, PROMALS, RiTO, RUTYLO, RUTYMETA, TD-ULPGC, TTM, varidis.
	Heuristic methods. Genetic algorithms. DDA, DECYL, Funaphy, GIOS, GOMA, GSO, Orel, PROMALS.
	Metaheuristic methods. DECYL, GIOS, LOGRO, RUTYLO, RUTYMETA.
	Game theory. CAG, GEUVA, GOR, GPB97, INFERES, MCS-UAB, MODES, modestya, RiTO, TD-ULPGC.
Biostatistics	Epidemiology. Clinical trials and treatment effectiveness. Genomics. DECYL, DEPREN, EE, GEUVA, GPB97, GRASS, GRID[ECMB], INFERES, MCS-UAB, MODES, MODESI, MODESMAN, modestya, OEDgroup, RiTO, TTM.
Geostatistics	Hydrogeology, oceanography, etc. CODA, GMFN, INFERES, MODESI, MODESMAN, modestya.

Table 3.4   Research groups in the Consulting Platform with relevant transfer experience in OTM

Category	Subcategory Groups
General	Resolution of systems of equations. CAG, DDA, EDANYA, MOSISOLID. Dynamical systems. DEPREN, GSD, HYPCHAOP, SSD. Graphs. CG, GAUCA, GIOS, RUTYLO, RUTYMETA, varidis.
Digital imaging	Processing, restoration and analysis of digital images. DEPREN, GSCUPM, HYPCHAOP, MODESMAN, TAMI.
Geometric analysis	Dynamic geometry. ACEIA. Intersection problems of geometric entities. CAG. Surface modelling. Surface reconstruction. TAPO. 3D reconstruction and visualization. EDANYA, MCS-UAB, TAMI.
Digital signal processing	Digital signal processing. Funaphy, GMFN, GSCUPM, HYPCHAOP.
GIS/GPS	Geographical information systems. CAG, GIOPTIM, MODESMAN, TD-ULPGC.
Communication networks	Telecommunication systems. Design of networks. GOMA, GSC, GSCUPM.
Codification of information	Cryptography. Codification and encryption algorithms. CAG, CG. Electronic security. CG, RiTO. e-voting. CG.
Computation	High performance computing. ADF, GIO, M2NICA, mat+i. Vector and parallel programming. EDANYA, GIO, modestya.

continued…

Category	Subcategory Groups
	GPU programming. EDANYA, M2NICA.
	Symbolic computation. Symbolic-numerical algorithms. ACEIA, CAG, EDnL, modsol, Psycotrip, SSD.
	Software package development. ACEIA, ADF, CAG, CODA, CYOPT, DDA, DECYL, EDANYA, EDnL, EE, EOPT, GMFN, GNOM, GOMA, GPB97, GRASS, GRID[ECMB], GSC, GSO, HYPCHAOP, INFERES, InterTech, KINETIC, M2NICA, M2S2M, MAI, mat+i, MCS-UAB, MODES, MODESI, MODESMAN, modestya, MOSISOLID, OEDgroup, PROMALS, Psycotrip, RUTYLO, RUTYMETA, TAMI, TAPO, TTM, varidis.
e-learning	Development and advanced use of e-learning tools. INFERES.
Searching & processing of information	Information diffusion algorithms. varidis.
Bioinformatics & Biomathematics	Bioinformatics. Biomathematics. Biomechanics. Physiology. Synthetic Biology. DEPREN, EDnL, GEUVA, GRASS, HYPCHAOP, INFERES, InterTech, MODESMAN, MOSISOLID, OEDgroup, TAMI, TAPO, TD-ULPGC.
Geomathematics	Geophysics. Geothermics. CAG, GMFN, InterTech, mat+i, MCS-UAB.
Optics	Optics. Photonics. InterTech, MCS-UAB, TAPO.

3.3.1
MSC classification

To sort the research activity of each group according to a classification scheme widely used by the research community, they have been categorised according to the 2010 version of the Mathematics Subject Classification (MSC2010).

Table 3.5 lists the MSC areas relating to the research activity of the groups in the Consulting Platform. For each MSC area, the groups which have participated in research projects, contracts, training or other transfer activities related to this area, are also included. It should be noted that our database contains information about related MSC research areas for only 64% of projects, 44% of contracts and 40% of training courses. For this reason, the information presented in this section is not as complete and exhaustive as we would have liked, since it was taken from only around 52% of the transfer activities developed by the research groups.

It is observed that most of the groups that have taken part in this study have developed transfer activities classified into the following MSC areas: *Operations research and mathematical programming* (90-xx codes), *Statistics* (62-xx codes), and *Numerical analysis* (65-xx codes), as was highlighted in Fig. 3.4 (see Sect. 3.1.2).

Table 3.5 First level MSC areas (code and area) and research groups in the Consulting Platform with transfer experience in them. Note: information about the MSC areas involved was only available for 52% of the transfer activities developed by the research groups

Code	MSC area Groups
03-xx	Mathematical logic and foundations LOGRO.
05-xx	Combinatorics LOGRO.
11-xx	Number theory CG, GAUCA.
14-xx	Algebraic geometry ACEIA, CG, GAUCA.
15-xx	Linear and multilinear algebra; matrix theory AALN, HYPCHAOP, MODESI.
16-xx	Associative rings and algebras GAUCA.
18-xx	Category theory; homological algebra MODESMAN.
31-xx	Potential theory varidis.
33-xx	Special functions InterTech.
34-xx	Ordinary differential equations M2NICA, SSD.
35-xx	Partial differential equations ADF, CYOPT, EDANYA, EOPT, GSC, M2NICA, M2S2M, MAI, mat+i, MOSISOLID.
37-xx	Dynamical systems and ergodic theory GSD, SSD.
39-xx	Difference and functional equations EOPT, MODESI.
42-xx	Harmonic analysis on Euclidean spaces Funaphy.
47-xx	Operator theory M2NICA, TOREFA.

continued...

Code	MSC area Groups
49-xx	Calculus of variations and optimal control; optimization ADF, CYOPT, Funaphy, GSC.
60-xx	Probability theory and stochastic processes DATFUN, DECYL, EOPT, MODESI, MODESMAN.
62-xx	Statistics ACEIA, CODA, DATFUN, DECYL, EE, EOPT, GEUVA, GIOPTIM, GIOS, GNOM, GPB97, GRASS, INFERES, MODES, MODESI, MODESMAN, OEDgroup, RUTYMETA, TTM.
65-xx	Numerical analysis AALN, CAG, CYOPT, DDA, EDANYA, Funaphy, GSC, GSD, HYPCHAOP, InterTech, M2NICA, M2S2M, MAI, mat+i, MODES, MOSISOLID, TAPO.
68-xx	Computer science CAG, EOPT, GIO, GOMA, HYPCHAOP, LOGRO, M2NICA, MODESI, Psycotrip, RUTYMETA, TAMI.
70-xx	Mechanics of particles and systems M2NICA.
74-xx	Mechanics of deformable solids M2NICA, MAI, mat+i, MODESI, MOSISOLID.
76-xx	Fluid mechanics EDANYA, EOPT, GSC, M2NICA, M2S2M, MAI, mat+i.
78-xx	Optics, electromagnetic theory InterTech, mat+i, TAPO.
80-xx	Classical thermodynamics, heat transfer InterTech, M2NICA, MAI, mat+i.
81-xx	Quantum theory InterTech, mat+i.
85-xx	Astronomy and astrophysics M2S2M.
86-xx	Geophysics CAG, M2NICA, M2S2M, MAI.
90-xx	Operations research and mathematical programming CAG, CYOPT, DECYL, EOPT, GIO, GIOPTIM, GIOS, GNOM, GOMA, GOR, GPB97, GSD, GSO, LOGRO, M2NICA, mat+i, MODES, MODESI, PROMALS, RUTYLO, RUTYMETA, TD-ULPGC, TTM.
91-xx	Game theory, economics, social and behavioural sciences CAG, GEUVA, GOR, M2NICA, mat+i, MODES, TD-ULPGC.
92-xx	Biology and other natural sciences DEPREN, InterTech, M2S2M, MOSISOLID, OEDgroup.

continued…

Code	MSC area Groups
93-xx	Systems theory; control AALN, CYOPT, GSC.
94-xx	Information and communication, circuits AALN, TAMI.
97-xx	Mathematics education EOPT, InterTech, M2NICA.

For each of the previous MSC areas, Fig. 3.5 shows the distribution of projects, research contracts, and training according to MSC classification. It should be highlighted that each of them can be associated to several MSC areas simultaneously. In order to simplify the chart, only MSC areas where more than 10 transfer activities had been conducted are presented (the figures for the remainder of the research areas were added in the *Other areas* category). Again, the MSC areas of *Statistics* (62-xx), *Operations Research and Mathematical Programming* (90-xx), and *Numerical Analysis* (65-xx) are highlighted, since most transfer activities carried out are partially or entirely related to them.

3.4
Expertise by sector: consulting and training

Mathematics provides a fundamental toolkit and a universal framework for innovation. The cross-cutting nature of mathematical and statistical techniques has favoured its interaction with a large number of industrial sectors. This section shows the extensive experience in knowledge transfer broken down by sector of economic activity for all research groups that took part in the survey. Information is presented both on an aggregate basis and itemised by research group. For each sector, a list of companies and industrial organisations which established partnerships for transferring knowledge or technology from those groups is included.

The data contained in this section is related to the ten sectors previously identified in Sect. 1.3. Each group of aggregate data refers to one of these main sectors of application. It is noteworthy that the full versions of the map have identified a total of 23 economic sectors for transferring mathematical techniques. These sectors were selected from previous reports (TransMATH Supply Map 2007, 2008, 2009 and 2010 editions), and later summarised for the preparation of this book. To do this, some areas have been created adding several economic sectors and others, considering both their lack of presence in the study of the demand and the lack of supply in knowledge transfer by groups for these sectors, have been either discarded (*Defence, Heritage Management & Conservation,* and *Livestock*), or instead included in one of the ten

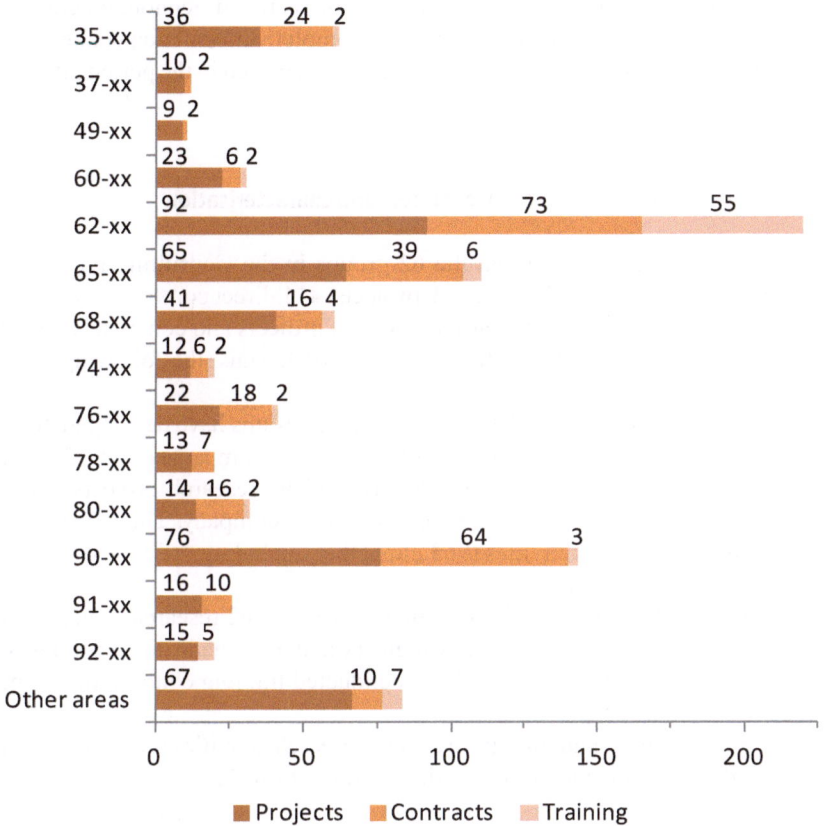

Fig. 3.5 Number of projects, contracts and training activities developed by research groups in the Consulting Platform by MSC area. Note: each project/contract/training course can be associated to several MSC areas simultaneously; moreover, information about involved MSC areas was only available for 64% of projects, 44% of contracts and 40% of training courses

sectors considered (*Agriculture, Marine Resources & Aquaculture*, and *Space*). In this last case, the consulting and training experience of the research groups was relocated as follows: knowledge transfer activities in *Agriculture* were distributed among *Biomedicine & Health, Energy & Environment, Food*, and *Logistics & Transport* sectors, depending on each activity and the type of companies involved; contracts and training courses in *Marine Resources & Aquaculture* were included in *Food*; and *Space* groups' experience now appears within the *Metal & Machinery* group.

Before putting forward the research groups' expertise for each sector of activity, a general overview of this experience in knowledge transfer in recent years (specifically, the period 2000–2010) is presented below, as well as a summarised review of the most recent research activities developed by the various groups.

Later, a graphical representation of global experience versus supply is shown, and the complete list of contracts and training courses in which the research groups have participated can be found broken down for each sector of economic activity. It is worth noting that these last items compile the complete information contained in the Consulting Platform database, that is, they are not restricted to the period 2000–2010.

3.4.1
Research activity from 2000 to 2010: overview and characterisation

Firstly, it must be highlighted that the 62 groups in the Consulting Platform have participated in 514 competitive research projects, 411 direct contracts with firms and 173 training courses. From them, around 90% of projects and contracts, and 70% of training courses were developed from 2000 to 2010. Thus, the following overview is focused on this period.

Restricting the analysis to 2000–2010, research groups have developed 467 competitive research projects, 375 direct contracts with companies and 121 training courses. In addition, it is noteworthy that 103 of the research projects developed (about 22%) have been supported in some way by a company. The global distribution of all these research activities (963 altogether, including projects, contracts and training) is shown in Fig. 3.6.

Furthermore, it must be emphasised that 40% of the 62 research groups have developed research projects supported by industry and 63% made direct contracts with companies, whereas 35% of groups have conducted training courses for companies during the period 2000–2010.

Since both contracts and training are those research activities most aimed at transferring mathematical technology to industry, the following subsections are devoted solely to them.

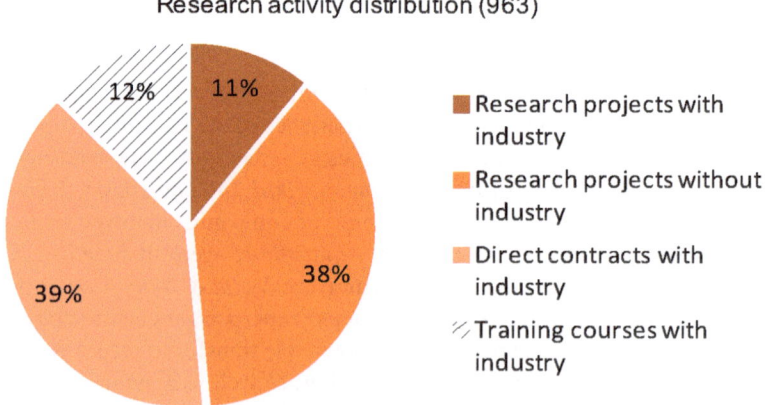

Fig. 3.6   Research activity distribution for research groups in the Consulting Platform during the period 2000–2010

3.4.1.1
Consulting and training (2000–2010): participants and duration

The 375 direct contracts with industry were developed by 39 of the research groups in the Consulting Platform (63%) for 197 different companies. The average duration of these contracts was 387 days. In fact, 34% of contracts had durations of more than a year.

As regards the training courses, it is noteworthy that 1,440 people from 70 different firms assisted them, their average duration being 27 hours. Moreover, 22 groups from the Consulting Platform (35%) developed this type of transfer activity during the period 2000–2010.

3.4.1.2
Consulting and training (2000–2010): by sector and research group

Figure 3.7 shows the total number of contracts and training courses for each sector during the period 2000–2010. Note that some of these knowledge transfer activities are associated simultaneously to several sectors of economic activity. Therefore, the sum of both contracts (474) and training courses (161) is higher than the values that we previously mentioned (375 contracts and 121 courses).

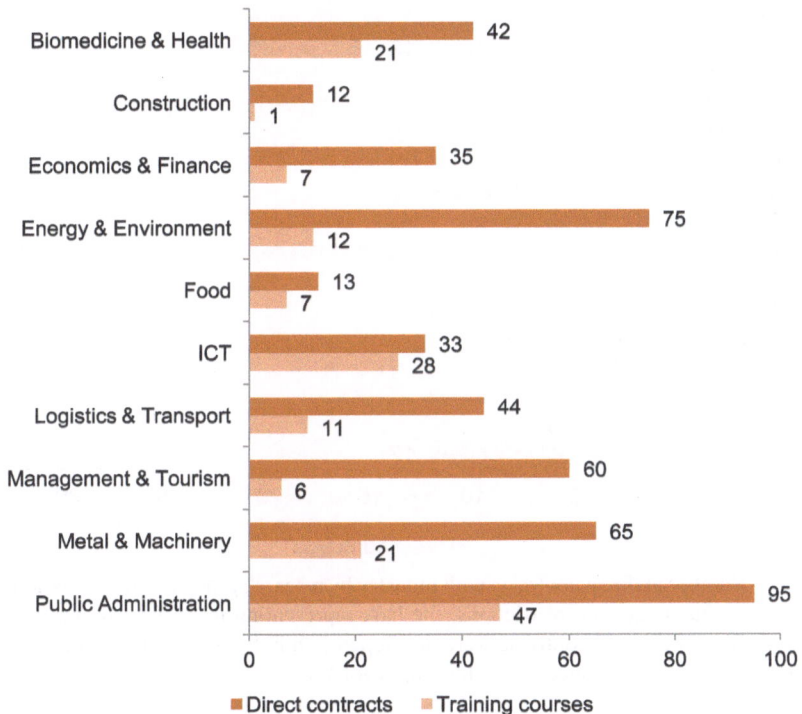

Fig. 3.7 Number of contracts and training courses by sector during the period 2000–2010. Note: some contracts/courses are associated to several sectors simultaneously

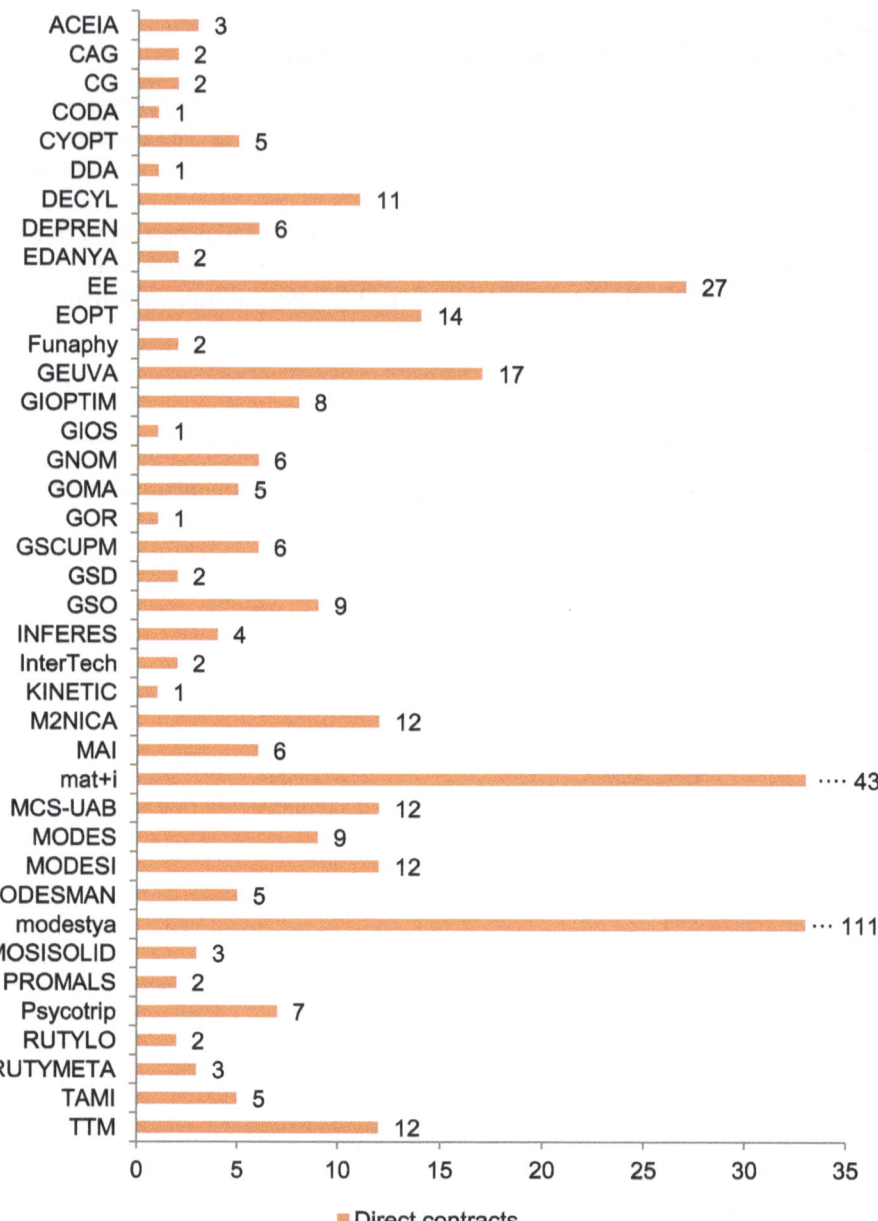

Fig. 3.8 Number of direct contracts with industry by research group during the period 2000–2010. Note: Due to scaling restrictions, the bars representing modestya and mat+i are not proportional, moreover, 7 contracts were developed jointly by two research groups, so they have been counted twice (once for each group involved)

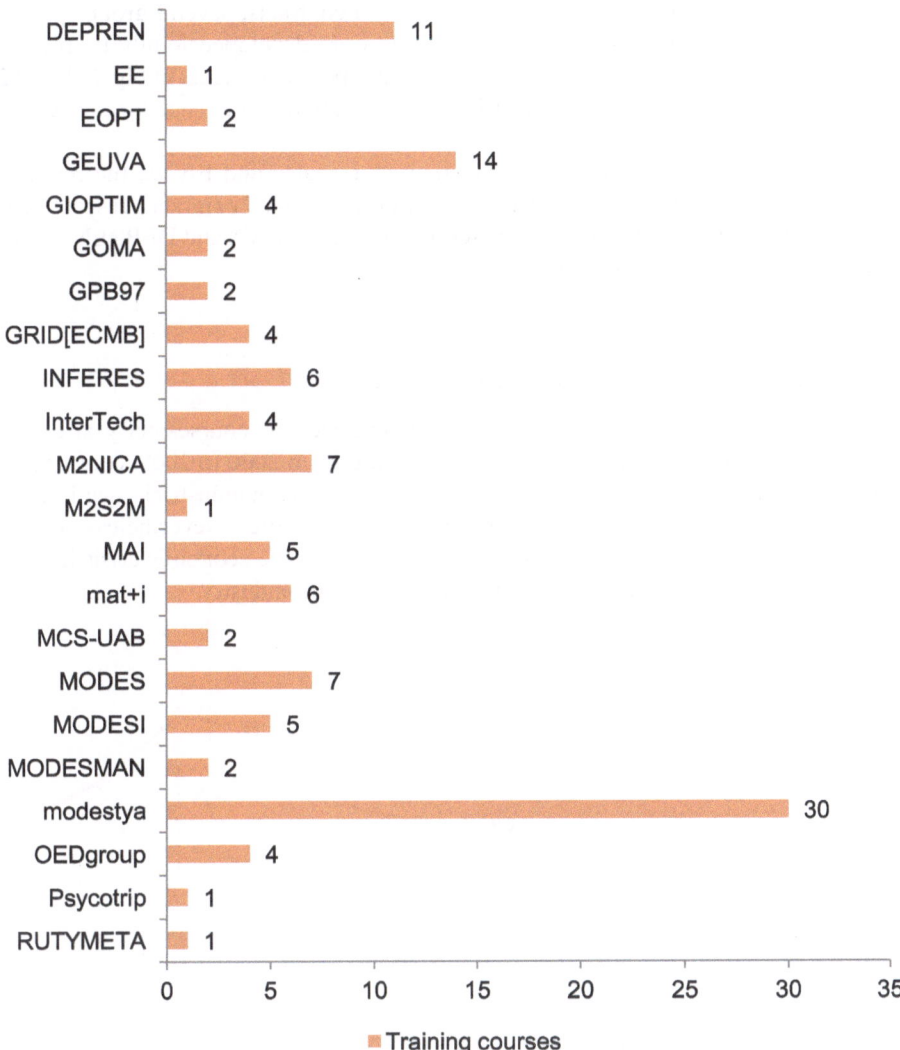

Fig. 3.9 Number of training courses by research group during the period 2000–2010

It is important to also note that most contracts and training courses carried out in that period are aimed at the *Public Administration* sector, with more than 90 contracts and 45 courses, the *Energy & Environment* sector, including 75 contracts and 12 courses, and the *Metal & Machinery* sector, with a figure of 65 contracts and over 20 training courses.

The distribution of the number of contracts and courses carried out in the period 2000–2010 for each research group is presented in Fig. 3.8 and Fig. 3.9 respectively. In all, 39 groups have carried out direct contracts, whereas 22 groups have participated in training courses with companies and industrial organisations within the period in question. We should state that, in some contracts, several groups participated simultaneously. Specifically, 7 contracts were developed jointly by two research groups. Due to this fact, the sum of the number of contracts in Fig. 3.8 is 382 instead of 375, since these 7 contracts have been counted twice (once for each group involved).

In terms of contracts with industry, modestya, mat+i and EE are the research groups which present the highest values, with 111, 43 and 27 contracts, respectively. Regarding the number of training courses, modestya, GEUVA and DEPREN are the most experienced research groups.

3.4.1.3
Consulting and training (2000–2010): time evolution

Figure 3.10 represents the number of current contracts and courses per year during the period 2000–2010. The trend is clearly upward from 2000 to 2008, corresponding to a closer engagement observed in recent years between industrial organisations and mathematical research conducted in Spanish universities. Nevertheless, the figures have significantly decreased since the beginning of the economic crisis in 2008 affecting, above all, the number of direct contracts with industry.

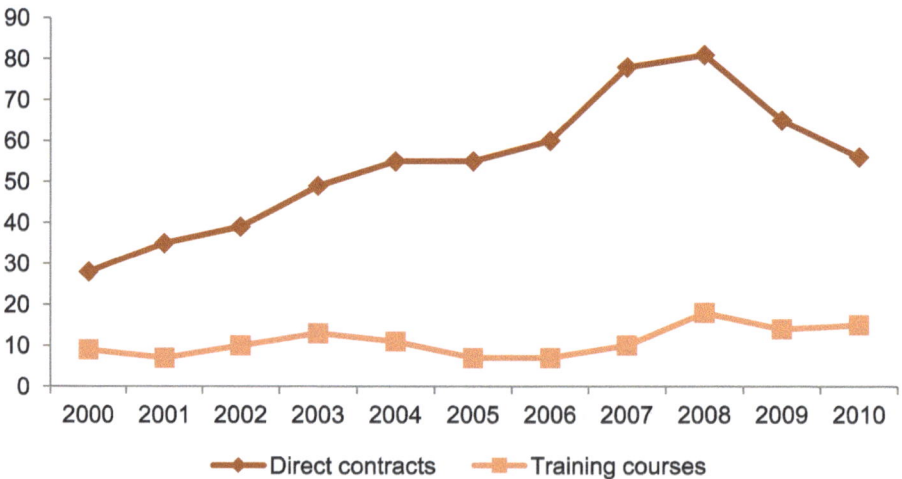

Fig. 3.10 Annual distribution of number of current contracts and training courses during the period 2000–2010

3.4.2
Technological transfer: supply and experience

A total of 50 of the 62 research groups (81%) have proven experience in knowledge and technology transfer to industry by means of research projects with industry, direct contracts, training courses and other knowledge transfer activities. It should be noted that, although the experience in knowledge transfer to industry is not particularly high for most groups, the values are significantly increased if considering the potential supply, as shown in Table 3.6 below, which represents both experience and provision, differentiated by sector. They show that the supply capacity of all groups significantly exceeds their proven experience.

As we have just mentioned, Table 3.6 summarises expertise and the current supply on offer for each group by sector. Cells marked with "x" refer to sectors in which groups offer technological services and have proven experience (whether in projects with industry, contracts, training or other knowledge transfer activities), whereas cells marked with "o" show those sectors where groups offer technological services but do not currently have expertise available.

Following this, information related to contracts and training courses developed by the research groups in the Consulting Platform is detailed by sector of economic activity. We should remark that each contract or course was assigned to just one main sector (that with which the contract or course is most related), with secondary related sectors also indicated. For each sector, the main applications of mathematical technology in this field are described. Subsequently, the lists of clients and research groups involved in the development of contracts and training courses, as well as the titles of those contracts and courses developed within the main sector of interest, are presented. It is worth highlighting that in distributing activities by region, 51% were carried out by Galician research groups (198 contracts and 97 training courses), 8% by groups from Castile and León (27 contracts and 17 training courses), 7% by groups from Navarra (41 contracts and 1 training courses), 6% by groups from Valencia (18 contracts and 17 training courses), and 5% by Catalonian research groups (27 contracts and 3 training courses).

Table 3.6 Supply and experience by research group and sector. "x" corresponds to groups with proven experience (projects with industry, contracts, training or other transfer activities) which offer technological services, while "o" shows those groups which are offering technological services in the sector, but do not currently have expertise available. The last row indicates the distribution of global technological supply by sector

	Biomedicine & Health	Construction	Economics & Finance	Energy & Environment	Food	ICT	Logistics & Transport	Management & Tourism	Metal & Machinery	Public Administration
ACEIA	o					x	x		x	x
ADF		o		o		x	o		o	
CAG						x	x	o	x	x
CG						x	o			o
CODA	x					x				x
CYOPT			o	x		x	x		x	o
DDA				x		x				o
DECYL	x			x		x	x	x		
DEPREN	x					x				x
EDANYA		x		x		x				o
EDnL	o		o			o				o
EE	x		x	x		x			x	x
EOPT			x	x	x	x	x			x
Funaphy		x							x	
GEUVA	x	o	x	x	x	o	x	x	x	x
GIO			o			o	o			o
GIOPTIM	x		x	x		x	x			x
GIOS				x			x			
GMFN		o		x		x				o
GNOM				x		x				x
GOMA						x	x			x
GOR			o			x	o			x
GPB97	x					x	x			x

continued…

	Biomedicine & Health	Construction	Economics & Finance	Energy & Environment	Food	ICT	Logistics & Transport	Management & Tourism	Metal & Machinery	Public Administration
GRASS	X	O			O	X				O
GRID[ECMB]	X				X	X				
GSC				O	O	X				
GSCUPM				X		X				
GSD	X		X					X		
GSO	X	O		X	O	X	X		O	X
HYPCHAOP	O	O				X			O	O
INFERES	X		X	X	X	X	X	O		X
InterTech	X			X		X			X	
KINETIC						X			X	
LOGRO		O	O	O		O	O	O		
M2NICA		X	X	X		X			X	
M2S2M				X		X			X	
MAI		X		X		X			X	
mat+i	O	X	X	X	X	X	X	X	X	X
MCS-UAB	X		X	X	X	X	X	X	X	X
MODES	X		X	X	O	X		O	X	X
MODESI	X		X	X	X	X	X	O	X	X
MODESMAN	X	X	O	X		X	X	X	X	X
modestya	X		X	X	X	X	X	X	X	X
MOSISOLID	O	O				X			X	X
OEDgroup	X			X	X	X		O		
Orel							O			
PROMALS						X	X	X		O
Psycotrip						X				
RiTO			O	O		O	O			O
RUTYLO		X				X	X		X	

continued…

	Biomedicine & Health	Construction	Economics & Finance	Energy & Environment	Food	ICT	Logistics & Transport	Management & Tourism	Metal & Machinery	Public Administration
RUTYMETA						x	x		x	
TAMI	o					x			x	
TAPO	x					x				
TD-ULPGC	o		o	o			o	o		o
TTM	o			x	x	x	x		x	o
varidis	o	o		o		x			o	
Supply (%)	11%	6%	8%	12%	5%	19%	11%	6%	10%	12%

3.4.3
Biomedicine & Health

Biomedicine & Health represents 11% of total supply from the groups and also 10% of proven experience in terms of contracts and training. The research groups have developed 28 contracts and conducted 33 training courses for this sector sector (reaching a total of 40 contracts and 39 training courses if the secondary sector *Biomedicine & Health* is also included), mostly related to data analysis and statistical studies for hospitals and health policy-makers. Among the applications in which the groups have remarkable expertise, we can highlight the following:

- analysis and design of experiments;
- design, development and analysis of surveys;
- studies of efficacy and safety of treatments;
- statistical analysis in epidemiology;
- statistical analysis of the effects of the use of drugs;
- tables of life expectancy;
- modelling of mortality tables;
- characterisation/grading of pharmaceutical and medical waste;
- processing and analysis of digital images; 3D visualization;
- applications in electronic engineering of medical devices;
- biomechanics; bones formation;
- numerical simulation of fractures, dental implants and orthodontic brackets;
- bioinformatics;
- mathematical biology; synthetic biology;
- computational design of proteins;
- metabolism; Arrhenius, compartmental and Michaelis Menten models;
- physiology of exercise;

- neurophysiologic modelling (evoked potentials);
- optimization and control of hospital processes;
- quality control in health care;
- location of services and health emergencies;
- shift allocation for health service employees;
- impact of geographical dispersion and population aging on healthcare spending;
- shift timetable scheduling;
- application and use of databases; data mining;
- statistical consultancy;
- electronic security systems.

3.4.3.1
Clients

- ASOCIACIÓN DE OCEANÓGRAFOS DE GALICIA
- ATOS ORIGIN
- BIOCROSS
- CIBER EPIDEMIOLOGÍA Y SALUD PÚBLICA
- CLUSTER DA ACUICULTURA DE GALICIA
- COLEGIO OFICIAL DE MÉDICOS DE CASTELLÓN
- COLEXIO OFICIAL DE MÉDICOS DA PROVINCIA DA CORUÑA
- COMPLEJO HOSPITALARIO JUAN CANALEJO
- COMPLEXO HOSPITALARIO ARQUITECTO MARCIDE/ PROF. NOVOA SANTOS
- CONSEJO SUPERIOR DE INVESTIGACIONES CIENTÍFICAS (CSIC)
- CONSELLERÍA DE SANIDADE (XUNTA DE GALICIA)
- FUNDACIÓN INSTITUTO GERONTOLÓGICO MATIA (INGEMA)
- FUNDACIÓN PÚBLICA ESCOLA GALEGA DE ADMINISTRACIÓN SANITARIA
- FUNDACIÓN VALENCIANA DE NEUMOLOGÍA
- GENENTECH ESPAÑA
- GEOGRAFÍA APLICADA, S.L.
- HOSPITAL CLÍNICO DE SALAMANCA
- HOSPITAL CLÍNICO UNIVERSITARIO LOZANO BLESA
- HOSPITAL GENERAL DE CASTELLÓN
- HOSPITAL GRAN VÍA DE CASTELLÓN
- HOSPITAL RIO HORTEGA DE VALLADOLID
- IDIS
- INSTITUTO DE SALUD CARLOS III DE MADRID
- INSTITUTO NAVARRO DE SALUD LABORAL
- INTUR and INFUR
- INVERNESS MEDICAL IBERICA
- JOHNSON & JOHNSON
- JUNTA DE CASTILLA Y LEÓN
- LABORATORIOS ISDIN
- MERCK & SHARP DOHME DE MADRID
- MUTUA NAVARRA
- NOVOTEC CONSULTORES
- PHARMA MAR
- PROTEOMIKA
- SERVEI DE ESTADÍSTICA – UNIVERSITAT AUTÒNOMA DE BARCELONA
- SERVICIO GALEGO DE SAÚDE (SERGAS – XUNTA DE GALICIA)
- SUBDIRECCIÓN XERAL DE INFORMACIÓN E DE SERVIZOS TECNOLÓXICOS
- UNIVERSIDADE DE SANTIAGO DE COMPOSTELA
- UNIVERSIDADE DE VIGO

3.4.3.2
Research groups

Groups with experience in the following section are:
DECYL, DEPREN, EE, GEUVA, GIOPTIM, GRID[ECMB], GSD, GSO, IN-
FERES, InterTech, MCS-UAB, MODES, MODESI, MODESMAN, modestya,
OEDgroup.

3.4.3.3
Consulting expertise

DECYL	• Analysis of the effectiveness of non-pharmacological treatments for Alzheimer's disease.
DEPREN	• Evaluation of the visual perceptual capabilities of athletes. Also of interest to the *ICT* sector. • First Olympic professorship. Also of interest to the *ICT* sector.
EE	• Statistical modelling for mortality risk in Navarra according to occupation.
GEUVA	• Collaboration in the development of a new laboratory test for in vitro diagnosis of Alzheimer's disease. • Cooperation agreement between the Consejería de Familia e Igualdad de Oportunidades (Department for Families and Equal Opportunities) of the Junta de Castile and León (regional government) and the University of Valladolid for the statistical support of the regional drugs commissioner. Also of interest to the *Management & Tourism* and *Public Administration* sectors. • Development of a project to generate classification algorithms for biological samples.
GIOPTIM	• Mathematical and statistical modelling for the development of geographic information systems in health. Also of interest to the *Public Administration* sector.
GSD	• Implementation of the Levenberg-Marquardt method for the least-squares fitting of nonlinear functions. *Jointly with MCS-UAB*.
GSO	• Application of optimization models in the location of health services in Castile and León. Also of interest to the *Public Administration* sector. • Development of models for locating emergency medical services in the region of Castile and León. Also of interest to the *Public Administration* sector.
INFERES	• Statistical software for the analysis of epidemiological studies and clinical experiments.

continued...

MCS-UAB	• Implementation of the Levenberg-Marquandt method for least-squares fitting of nonlinear functions. *Jointly with GSD*. • Report on the effectiveness and safety of acne treatment with Niacex (4% Niacinamide).
MODESI	• Statistical analysis of quality-related questionnaires. • Statistical treatment of the study of research: "Risk factors associated with muscular-skeletal deterioration of the upper body and the spinal column".
MODESMAN	• Physical activity and EPOC.
modestya	• Design of a model to show the impact of geographic dispersion and population aging on healthcare spending. • Modelling and prediction of time in surgical waiting lists. Also of interest to the *ICT* sector. • Risk factors for hospital readmission of people of 75 and over: development of a prediction model using CMBD. • Statistical study of the characterisation of pharmaceutical waste in the SIGRE system (collection and recycling of unwanted medicines). Also of interest to the *Energy & Environment* sector. • Statistical study for the valuation of the clinical effectiveness of a new pharmaceutical product for the treatment of the *philasterides dicentrarchi* parasite in turbot. • Technological support and development of training activities in statistical data analysis applied to Epidemiology (I–VII).

3.4.3.4
Training expertise

GRID[ECMB]	• Biostatistics course using R software. • Flexible regression by means of generalized additive models (GAM). Biomedical applications using R. • Penalised spline regression and geoadditive regression using BayesX. Also of interest to the *Economics & Finance, Energy & Environment, Food* and *Management & Tourism* sectors.
INFERES	• CSIC research based training course (I-II). • Applied course in SPSS and R.
InterTech	• Introduction to Synthetic Biology (I-II).
MODES	• Data analysis with SPSS software (I-V). • Experimental design and analysis. • Survival analysis applied to R.
MODESI	• The application of Statistics in research.

continued…

MODESMAN	• The application of Statistics to the publication of scientific projects: use of the SPSS software. • The application of Statistics to the publicity of scientific projects: use of the SPSS software. • Medical informatics: handling statistical data with SPSSWIN. • S-PLUS and its statistical applications. • Course in statistical analysis using computers: SPSS/PC+ package. • Postgraduate course in Statistics for Epidemiology by S-PLUS.
modestya	• Analysis of variance. • Biostatistics. • Introduction to the SPSS statistical package (I-II). • Logistic regression. • Practical application of the SPSS statistical package. • Research methodology. Data analysis (handling the SPSS package). • Survival analysis. • Time series analysis.
OEDgroup	• Data analysis (I-II).

3.4.4
Construction

There are a total of 16 of the 62 research groups (6%) currently offering supply and technical services to the *Construction* sector. Furthermore, 6 of them have proven expertise in making contracts in this field, which represents the sector where the least collaborations have been made (2% of total). Specifically, the research groups have developed 10 contracts in this sector (reaching a total of 14 contracts and 1 training course if the secondary sector *Construction* is also included). Below, some applications for mathematical technology in *Construction* are listed:

- simulation of buildings' thermal and acoustic insulation;
- simulation of noise reduction barriers;
- numerical simulation of load tests on bridges;
- vibration of structures;
- numerical characterisation of insulating, lightweight and thermal materials;
- numerical characterisation of resistance of materials;
- numerical calculation of transfer coefficients according to ISO standards;
- numerical simulation of ventilated façades;
- numerical simulation of fire;
- statistical assessment for quality control;
- study of the suitability of soil in construction;
- buildings' resistance to climatic effects;
- building durability;
- calculation of CO_2 emissions in cement factories;
- software for optimization of stone cutting.

3.4.4.1
Clients

- AT CONTROL
- COLEGIO OFICIAL DE ARQUITECTOS DE VA-LENCIA
- CONSELLERÍA DE INNOVACIÓN E INDUSTRIA (XUNTA DE GALICIA
- FUNDACIÓN CESGA

- GRANITOS MONTE FARO
- INDUSTRIAS GONZÁLEZ
- ITENE
- SEÑALIZACIONES POSTIGO
- TALLERES CORTES

3.4.4.2
Research groups

Groups with experience in the following section are:
Funaphy, M2NICA, MAI, mat+i, MODESMAN, RUTYLO.

3.4.4.3
Consulting expertise

Funaphy	• Framework collaboration agreement between the Señalizaciones Postigo S.A. company and the Technical University of Valencia. Also of interest to the *Metal & Machinery* sector.
M2NICA	• Free simulation software in companies. *Jointly with MAI and mat+i.* Also of interest to the *Metal & Machinery* sector.
MAI	• Free simulation software in companies. *Jointly with M2NICA and mat+i.* Also of interest to the *Metal & Machinery* sector.
mat+i	• Development of a computer application for the optimization of cutting granite blocks to meet an order, in accordance with the Montefaro cutting system. Also of interest to the *Logistics & Transport* sector. • Free simulation software in companies. *Jointly with M2NICA and MAI.* Also of interest to the *Metal & Machinery* sector. • Obtaining lightweight insulating materials for the building industry. Also of interest to the *Metal & Machinery* sector.
MODESMAN	• Automation of the manufacturing process and on-site storage of concrete test samples for quality control in buildings. • Statistical quality control of ceramic mouldings. • Statistical study of experimental work for the determination of the critical chloride content in samples of concrete made with aluminous cement. • Statistical study of experimental work for the determination of the resistance of concrete beams using microsamples.
RUTYLO	• Advanced software for the optimization of vehicle routing. Also of interest to the *ICT* sector. • Optimization modules for transport planning.

3.4.5
Economics & Finance

Economics & Finance represents 8% of the total supply from the 62 research groups in the Consulting Platform, with 7% of their proven expertise. Specifically, the latter has 29 contracts and 10 training courses in this sector (reaching a total of 32 contracts and 14 training courses if the secondary sector *Economics & Finance* is also included), focused on lines of work such as models for the development of financial products, risk management or analysis and prediction aimed at decision-making, in high demand by financial institutions and banks. Regarding mathematical applications in the *Economics & Finance* sector, we find:

- market studies;
- characterisation of dynamical trajectories according to market sectors;
- statistical and mathematical models of client populations;
- statistical analysis of socio-economic indicators according to territorial groups;
- Economic Geography;
- analysis and prediction of interest rates; Quantitative Finance;
- valuation of products and financial derivatives;
- estimation of economic variables according to industrial sectors;
- actuarial models of risk; risk prevention;
- stochastic modelling of catastrophe risks;
- cryptography; electronic security; data protection systems;
- studies of quality of services.

3.4.5.1
Clients

- ANALISTAS FINANCIEROS INTERNA-CIONALES (AFI)
- APLICACIONES DE INTELIGENCIA ARTIFICIAL (AIS)
- BRICOKING
- CAIXA CATALUNYA
- CAIXA GALICIA
- CAJA DE AHORROS DEL MEDITERRÁNEO
- CONSELLO DE CONTAS DE GALICIA
- ESCUELA DE FINANZAS DEL BBVA
- FUNDACIÓN CAIXA GALICIA – CLAUDIO SAN MARTÍN
- FUNDACIÓN PARA EL FOMENTO DE LA CALIDAD INDUSTRIAL Y EL DESARROLLO TECNOLÓGICO DE GALICIA
- INDUSTRIA DE DISEÑO TEXTIL (INDITEX)
- INSTITUT D'ANÀLISI ECONÒMICA
- INSTITUTO DE ESTUDIOS FISCALES
- LKS
- NFC Y SUEÑO
- NOVA CAIXA GALICIA
- PROMAINSUR
- SEGUROS CAJASUR
- TECNOLOGÍA DE INFORMACIÓN Y FINAN-ZAS
- TELECYL
- UNIVERSIDADE DE VIGO

3.4.5.2
Research groups

Groups with experience in the following section are:
EE, EOPT, GEUVA, GIOPTIM, GSD, INFERES, M2NICA, mat+i, MCS-UAB, MODES, MODESI, modestya.

3.4.5.3
Consulting expertise

EE	• Models of collective housing price forecasts in Álava. Also of interest to the *Management & Tourism* sector. • Statistical analysis and modelling of collective housing prices in Álava (I-II).
GEUVA	• Advice and collaboration in research and development projects of generic models of data mining (I-II).
GIOPTIM	• Development of a mathematical model for optimizing cash deposits in bank branches. Also of interest to the *ICT* and *Logistics & Transport* sectors. • Overhaul and update of models to assist decision-making for optimizing cash deposits in bank branches. Also of interest to the *ICT* and *Logistics & Transport* sectors.
GSD	• Design of an algorithm for the characterisation of dynamic trajectories in the housing market. • Development of an algorithm for the simulation of a system of differential equations applied to the housing market.
M2NICA	• Development of an application of optimization for insurance portfolios and risk-assessment. *Jointly with mat+i.* • Insurance company ALM GPU simulation. • Numerical resolution of models for the valuation of financial products (I-III). *Jointly with mat+i.* • Valuation and generation of portfolio flows. • Valuation model for autocallable financial products and with continuous barrier, and development of software for their valuation. • Valuation model for structured products and development of software for their valuation.
mat+i	• Development of an application of optimization for insurance portfolios and risk-assessment. *Jointly with M2NICA.* • Numerical resolution of models for the valuation of financial products (I-III). *Jointly with M2NICA.*
MCS-UAB	• Development of a code in Fortran for the study of auctions.

continued...

MODES	• Analysis and application of statistical techniques aimed at decision making within a business framework. • Analysis and development of stochastic-mathematical models for behaviour of sample client populations receiving credit loans. • Stochastic modelling of accident rates: the Inditex Group case.
MODESI	• Design of the sample for a survey into the consequences of shift work on employee health. Also of interest to the *Biomedicine & Health* sector. • Statistical analysis of the data from a survey on the consequences of shift work on employee health. Also of interest to the *Biomedicine & Health* sector.
modestya	• Analysis and prediction of prices of financial assets (I-II). • Analysis and prediction of the temporal structure of interest rates. • Identification of office groups according to socio-demographic and financial characteristics. • Operating risks: measure and management. • Studies for the analysis and knowledge of the evolution of the Galician economy.

3.4.5.4
Training expertise

EOPT	• Statistics for risk management.
GSD	• Asset valuation with MATLAB.
INFERES	• Introductory course on statistical techniques with SPSS (I-III). • Training course on statistical support tools for the functioning of technological surveillance. Also of interest to the *Logistics & Transport* sector.
M2NICA	• Graphics processors for financial calculations.
MCS-UAB	• Course on asset and liability management. Also of interest to the *Logistics & Transport* sector. • Mathematical course for financial instruments.
modestya	• Introduction to panel data.

3.4.6
Energy & Environment

Energy & Environment represents 12% of the total supply from the groups and 15% in terms of proven experience (being the second sector with the largest number of contracts and courses, after *Public Administration*). In the Consulting Platform database, 79 direct contracts with industry and 11 training courses were recorded by the research groups for this sector (reaching a total of 90 contracts and 20 training courses if the secondary sector *Energy & Environment* is also included). The width of the groups' expertise in solving energy and environmental problems is shown in the following list of research lines which have been developed:

- Computational Fluid Dynamics (CFD);
- heat transfer;
- numerical simulation of heat transfer and combustion processes;
- applications in energy installations such as numerical simulation of combustion in coal and oil boilers, prediction of breaks, pyrotechnic pasta combustion, characterisation of thermal groups, etc.;
- modelling and simulation of oxy-combustion in pulverized coal flames;
- Chemical Kinetics;
- statistical models for environment and analysis of spatial data;
- control and calculation of emissions;
- microgravimetric studies; Environmental Geophysics;
- waste characterisation;
- packaging content analysis in urban waste;
- control of processes;
- prevention and control of environmental pollution;
- hydro-electric resources; urban and marine hydro-electric power;
- hydrogeological models; modelling in aquifers;
- simulation of flooding;
- simulation and prediction of air and water quality;
- applications in Oceanography;
- modelling of fires;
- dispersal of pollutants;
- risk prevention;
- renewable energy;
- geostatistical modelling of climatic variables; weather maps;
- maps of wind and wind power potential studies; economic feasibility of wind farms;
- solar maps and solar radiation;
- geothermal acclimatisation; determination of thermal parameters of structures and subsoil in real time;
- simulation and control of acclimatisation systems based on heat pumps;
- production planning of hydrothermal energy;
- management models of mixed energy systems;
- energy storage and processing system feasibility;
- problem solving in plasma physics;
- production forecasting and planning;
- optimization of power distribution networks;
- calculation of electromagnetic parameters on underground power lines.

3.4.6.1
Clients

- ABENGOA SOLAR NEW TECHNOLOGIES
- AGROAMB
- AIGUASOL ENGINYERIA
- AIMEN
- ALCOA
- ANBIOTEK
- APLICACIONES TOPOGRÁFICAS Y CARTOGRÁFICAS
- APPLUS NORCONTROL
- AYUNTAMIENTO DE ZARAGOZA
- CEMENTOS PORTLAND VALDERRIVAS
- CENER-CIEMAT (CENTRO NACIONAL DE ENERGÍAS RENOVABLES)
- CENTRO DE PESQUISAS DE ENERGIA ELÉTRICA (CEPEL BRASIL)
- CESPA
- CONFEDERACIÓN HIDROGRÁFICA DEL EBRO
- CONSEJERÍA DE MEDIOAMBIENTE (GOBIERNO DE ARAGÓN)
- CONSULTEC INGENIEROS
- DESARROLLOS EÓLICOS (DESA)
- DIRECCIÓN GENERAL DE MINAS DEL PRINCIPADO DE ASTURIAS
- ECOEMBALAJES ESPAÑA
- EMPRESA FUNDACIÓN PARA EL FOMENTO DE LA CALIDAD INDUSTRIAL Y EL DESARROLLO TECNOLÓGICO DE GALICIA
- EMPRESA MUNICIPAL DE AGUAS DE A CORUÑA (EMALCSA)
- ENDESA
- ENDESA GENERACIÓN
- ENDESA SERVICIOS
- ENERGESIS INGENIERÍA
- ESRI
- EUROPIZARRAS
- EWATER CRC
- FACET IBÉRICA
- FERROATLÁNTICA
- FUNDACIÓ UNIVERSITAT-EMPRESA DE LA UNIVERSITAT DE VALÈNCIA (ADEIT)

- FUNDACIÓN AGUAS DE BARCELONA
- FUNDACIÓN CIUDAD DE LA ENERGÍA
- GEOMODELS
- GERENCIA REGIONAL DE SALUD DE CASTILLA Y LEÓN
- GESAN
- GESTIÓN AMBIENTAL DE VIVEROS
- IMPIVA
- INKOA
- INSTITUTO OCEANOGRÁFICO ESPAÑOL DE MADRID
- INSTITUTO TÉCNICO Y DE GESTIÓN AGRÍCOLA (ITGA)
- INTERNACIONAL GEOSERVICES
- INSTITUTO SISTEMAS FOTOVOLTAICOS DE CONCENTRACIÓN (ISFOC)
- LIGNITOS DE MEIRAMA
- NEXANS IBÉRICA
- NORCONTROL
- NORVENTO ENERXÍA
- ORONA S. COOP
- PRINCIPADO DE ASTURIAS
- RECICLAUTO
- REPOBLACIONES FORESTALES
- REPSOL EXPLORACIÓN
- RÍO NARCEA GOLD MINES
- SGL CARBON
- TECNALIA RESEARCH & INNOVATION
- TOTEMA ENGINEERING
- UFISA SOLUZIONA
- UNIÓN ESPAÑOLA DE EXPLOSIVOS
- UNIÓN FENOSA GENERACIÓN
- UNIVERSIDAD DE OVIEDO
- UNIVERSIDADE DE SANTIAGO DE COMPOSTELA
- UNIVERSITAT DE VALENCIA
- UNIVERSITAT POLITÈCNICA DE VALÈNCIA
- XUNTA DE GALICIA

3.4.6.2
Research groups

Groups with experience in the following section are:
CYOPT, DDA, DECYL, EDANYA, EE, EOPT, GEUVA, GIOPTIM, GIOS, GMFN, GNOM, GSCUPM, INFERES, InterTech, M2NICA, MAI, mat+i, MCS-UAB, MODES, MODESI, MODESMAN, modestya, OEDgroup, TTM.

3.4.6.3
Consulting expertise

CYOPT	• PLCBAS program license. • Software for the planning and optimization of the coating phase in packaging.
DDA	• Validation of software for three-dimensional wind farm simulation and its application in the study of wind power.
DECYL	• Assessment of a system for energy transformation and storage. • Dimensional analysis of a plant for decontamination of end-life-vehicles and transport logistics. Also of interest to the *Logistics & Transport* sector. • Logistics analysis of a glass recycling system. *Jointly with GIOS*. Also of interest to the *Logistics & Transport* sector. • Management models of hybrid wind-hydrogen energy systems. Improvement of WindHyGen software.
EDANYA	• Simulation of a torrent flood in the village of Maro, Malaga (I-II). Also of interest to the *Construction* sector.
EE	• Climate mapping geostatistical techniques. • European Foresee project technical-statistical assessment.
EOPT	• Calculation of CO_2 emissions from Cementos Pórtland Valderrivas factories. *Jointly with TTM*.
GEUVA	• Power technology development in the Castile and León Health System. Also of interest to the *Biomedicine & Health* and *Public Administration* sectors.
GIOPTIM	• Subcontracting agreement between Abengoa Solar New Technologies, S.A. and Research Foundation, University of Sevilla.
GIOS	• Logistics analysis of a glass recycling system. *Jointly with DECYL*. Also of interest to the *Logistics & Transport* sector.

continued…

GMFN	• Geostatistical modelling of climate variables. • Geostatistical study of Carlés gold deposit. • Geotechnical analysis of foundations by 2-D seismic tomography in transmission. • Microgravimetric study of the influence of the Llumeres Group mining activities.
GNOM	• License agreement of the PPRN package (only linear capabilities), including maintenance support.
GSCUPM	• Robustness in complex systems.
INFERES	• Sampling and analysis of domestic electric energy induction meters in Galicia. Also of interest to the *Logistics & Transport* sector.
InterTech	• Analysis and determination of the thermal conductivity of structures and subsoil in real time. Also of interest to the *Metal & Machinery* sector.
M2NICA	• Computer application for the calculation of electromagnetic parameters of overhead and subterranean power lines.
MAI	• Design of seismic acquisition parameters for the resolution of seismically opaque zones. • Modelling and design of a slate mine in Vilarchao. • Study of the combustion of pyrotechnic pastes. • Thermal analysis of heating ovens for the tempering of steel parts. Also of interest to the *Metal & Machinery* sector.
mat+i	• Application of numerical simulation technology in fluid mechanics (CFD) for the study of processes in energy-generating facilities. • Characterisation of the thermal generators at the As Pontes Thermal Power Plant. • Hydro-chemical evaluation of organic solute transport in the mining lake of Meirama. • Modelling and computation of fluid dynamics and combustion in the As Pontes TPP boiler. • Modelling and computation of fluid dynamics and combustion in the Maritsa-East 3 (Bulgaria) TPP boiler. • Modelling and simulation of oxycombustion in pulverized Carbon limpets (SIMU-LOX) (I-II). • Numerical simulation of the effect on temperature of the introduction of pulverized water into a chimney. Optimization of the position of the pulverizers. • Research and development of a detailed program for the closing of open-cast mines. • Research into "NET ZERO" (NETOLIFT) lifting technology. • Simulation of the furnace at the As Pontes Thermal Power Plant by computational fluid dynamics (CFD) (I-III). • Technical assistance in the application of numerical simulation technology in fluid mechanics (CFD) for the study of processes in energy-generating facilities (I-II).

continued...

MCS-UAB	• Building of a linear model to approximate simulation with TRANSOL. • Improvement and automation of algorithms for the 3D reconstruction of geological structures.
MODES	• Extension of statistical techniques for the control of stochastic and dynamic variables associated with the electrical operation of induction furnaces of the photovoltaic quality silicon project. • Metrological control of counters park of EMALCSA. • Statistical analysis of the monitoring of water and particulate content in aviation fuel. • Statistical assessment for the performance of sampling plans on the content and characterisation of the so-called "yellow pocket" waste in the Galician municipalities belonging to the Sogama Plan. • Statistical assessment of a quality control problem in the manufacturing process of graphite electrodes. Also of interest to the *Metal & Machinery* sector.
MODESI	• Analysis of seasonal trends in series of rainfall in the Ebro Valley. • Management indicators for water supply. • Study of the relationship between the generation of waste and economic activity in Aragón. • Tapping of resources in the Ebro basin. Regional analysis of the trends in seasonal rainfall.
MODESMAN	• Development of statistical data analysis of marked point patterns. Also of interest to the *Metal & Machinery* sector.
modestya	• Analysis of statistical variables related to the operation of the Thermal Power Station in La Robla (León). • Design and development of a mathematical model for cold bending pipeline over ten inches. • Design of an immission prediction system for the environs of the As Pontes U.P.T. by means of neuronal networks. • Diachronic valuation of use of sewage sludge in agriculture: production, biodiversity, phosphorus and heavy metals. • Improvement of immission prediction system (SIPEI 2003). • Improvements in the statistical prediction of immission by means of the implantation of probability prediction methods for episodes of contamination, and introduction of prediction with neuronal networks. • Introduction of statistics applied to the analysis of urban waste. • System for the Statistical Prediction of Immissions, SIPEI 2004 to 2011 (I-VIII). • Specification for the modernisation of an air quality prediction system and adaptation to new legislative requirements. • Statistical assessment for the characterisation of urban waste in Asturias. • Statistical assessment for the design of a methodology to determine the content of packaging in urban waste. • Statistics – neuronal networks for the Besós Thermal Energy Plant.

continued ...

OEDgroup	• Technical assistance contract.
TTM	• Calculation of CO2 emissions from Cementos Pórtland Valderrivas factories. *Jointly with EOPT*. • Development of an integrated control system for agroclimate variables in waste disposal. • Forecast of flexibility of electricity consumption according to price signals. • Methodology proposal for the determination of minimal ecological flow adapted to the natural hydrogram of a river. • Rapid formula for evaluating cleaning costs in a city. • Rapid formula for the evaluation of optimal refuse collection in a city.

3.4.6.4
Training expertise

GMFN	• Geostatistical modelling: applications in mining, hydrogeology and environment. • Mathematical models in hydrogeology. • Notes on geostatistical modelling.
InterTech	• Design of installations with geothermal heat pump. • Energy sustainability and integration of renewable energy in buildings.
MODES	• Basic statistics and experimental design. Also of interest to the *Metal & Machinery* sector. • Regression models. Also of interest to the *Metal & Machinery* sector.
MODESMAN	• Analysis of environmental data. • Analysis of statistical data with S-PLUS.
modestya	• Application of statistical analysis to atmospheric contamination control (II Master in Science, Technology and Environmental Management). • Prediction of plume transport using time series (III Seminar on non-parametric inference).

3.4.7
Food

A total of 14 groups supply technological services and training in the *Food* sector (5% of total supply). Moreover, contracts and training courses for this sector represent 3% of total proven experience, 10 groups having expertise related to this sector. In fact, the research groups have participated in 11 contracts and conducted 6 training courses for this sector (reaching a total 12 contracts and 11 training courses if the secondary sector *Food* is also included). Some applications, in which the research groups have experiencein technological transfer, are detailed below:

- market studies;
- sensory analysis;
- control and optimization of products and processes;
- provisioning optimization for distribution companies and large chains;
- production planning;
- allocation of products to customers;
- sterilisation of food;
- analysis and optimization of the sterilisation processes of canned food;
- shelf life of foods;
- evaluation of human exposure to chemicals inside food;
- epidemiological analysis of the influence of diet on health.

3.4.7.1
Clients

- AULA DE PRODUCTOS LÁCTEOS – UNIVERSIDADE DE SANTIAGO DE COMPOSTELA
- AZUCARERA DE VALLADOLID
- CENTRO DE COMPUTACIÓN ESPECIALIZADA
- CENTRO TECNOLÓGICO DE LA PESCA (CETPEC)
- CONSEJO SUPERIOR DE INVESTIGACIONES CIENTÍFICAS (CSIC)
- EROSKI
- GOBIERNO VASCO
- GRUPO DE RECURSOS MARINOS Y PESQUERÍAS – UNIVERSIDADE DA CORUÑA
- IMPULSO INDUSTRIAL ALTERNATIVO
- JAVIER SANGENÍS
- MERCAZARAGOZA
- PRODUCT SENSORY CONSULTING GROUP
- SCC
- SIDRERÍA GALLEGA
- STOLT SEA FARM
- SYNGENTA SEEDS
- UNILEVER FOOD ESPAÑA
- UNIVERSIDAD DEL PAÍS VASCO – EUSKAL HERRIKO UNIBERTSITATEA

3.4.7.2
Research groups
Groups with experience in the following section are:
EOPT, GEUVA, GRID[ECMB], INFERES, mat+i, MCS-UAB, MODESI, modestya, OEDgroup, TTM.

3.4.7.3
Consulting expertise

| EOPT | • Optimization of supply for a large retail chain. Also of interest to the *Logistics & Transport* sector. |
| | • Optimization of supply in distribution and delivery companies. Also of interest to the *Logistics & Transport* sector. |

continued...

GEUVA	• Validation and sensitivity analysis of a probabilistic model for assessment of dietary exposure to pesticide residues. Also of interest to the *Biomedicine & Health* and *Public Administration* sectors.
mat+i	• NORDÉS (Wind power propulsion on boats: design and simulation). • Optimization of the size of productive units for the design of large tonnage aquaculture plants.
MCS-UAB	• Generation of product matching to consumers.
MODESI	• Generation of numbers for the drawing of lots.
modestya	• Consumer study for Sidrería Gallega (Galician Cider Co.). • Sensorial analysis study for Sidrería Gallega. • Statistical data analysis and assessment in different research projects.
TTM	• Epidemiological consultancy in the study of cholesterol.

3.4.7.4
Training expertise

GRID[ECMB]	• Multiple regression models with R: GLM and GAM.
INFERES	• Statistics with R.
MODESI	• Masters in logistics and food safety.
modestya	• Experimental design and analysis of results.
OEDgroup	• Experimental design. • Statistical techniques of data analysis for research into horticulture.

3.4.8
ICT

Technology and knowledge transfer to the *ICT* sector represents 19% of the total supply from the 62 research groups, and 8% of their total expertise. As well as this experience, their work on the development of software within the framework of projects for other sectors is of great relevance, since mathematical technology is, in many cases, integrated into specific software packages. Specifically, groups in the Consulting Platform have developed 31 contracts and 18 training courses for the *ICT* sector (reaching a total 39 contracts and 33 training courses if the secondary sector *ICT* is also included). Some research lines in the *ICT* sector where mathematical techniques have successfully been used are the following:

- geographic information systems;
- processing, analysis and restoration of digital images;
- cryptography; computer and network security; data protection systems;
- algorithms for the dissemination of information;
- e-learning;
- management and planning of telecommunication networks;
- design of wireless networks;
- simulation of the behaviour of electronic devices;
- modelling of microstructured optical fibers; design of waveguides;
- scientific and technical advice on databases;
- software development solutions on demand using mathematical, statistical and numerical simulation;
- ICT assessment;
- programming on GPUs.

3.4.8.1
Clients

- ADDLINK RESEARCH
- ADDLINK SOFTWARE CIENTÍFICO
- AGENCIA DEL CONOCIMIENTO Y LA TEC-NOLOGÍA
- ALCOA-INESPAL
- ARSYS INTERNET
- CENTRO TECNOLÓGICO DE COMPONENTES
- COMERCIO ELECTRÓNICO B2B 2000
- CONSEJO SUPERIOR DE DEPORTES
- CONSELLERÍA DE CULTURA, COMUNI-CACIÓN SOCIAL Y TURISMO (XUNTA DE GALICIA)
- DXO LABS
- FIBA EUROPA

- HEPTAGON
- INDRA SISTEMAS
- INTEGRACIÓN Y CONTROL
- JUNTA DE EXTREMADURA
- M PUNTO 2 PUNTO TECHNOLOGIES
- NKT RESEARCH & INNOVATION A/S
- OMNIATEC
- OPTIMIZATION TECHNOLOGIES
- PROCEDIMIENTOS UNO
- SCYTL SECURE ELECTRONIC VOTING
- TELEFÓNICA I+D
- UNIVERSIDAD COMPLUTENSE DE MADRID
- UNIVERSITAT POLITÈCNICA DE VALENCIA
- VODAFONE ESPAÑA

3.4.8.2
Research groups

Groups with experience in the following section are:
ACEIA, CAG, CG, DECYL, DEPREN, EDANYA, GIOPTIM, GOMA, GSCUPM, HYPCHAOP, InterTech, M2NICA, MAI, mat+i, MODESI, MODESMAN, modestya, Psycotrip, RUTYMETA, TAMI.

3.4.8.3
Consulting expertise

ACEIA	• Educational software for secondary school mathematics teaching in free software algebraic computation systems.
CAG	• Development of a guidance calculation system in real time based on non-dedicated GPS's.
CG	• ECC algorithms for e-voting. • Generation of promotional codes.
DECYL	• Modelling the problem of bandwidth allocation in the transmission of information packets in All-IP communication systems with queuing theory.
DEPREN	• Development of a 3-D representation prototype to be used in virtual training. Also of interest to the *Biomedicine & Health* sector. • Technical support for Spanish sporting federations. • To evaluate the visual skills of basketball umpires. Also of interest to the *Biomedicine & Health* sector.
EDANYA	• Contract with the company Procedimientos UNO S.L. for the licensing and marketing of software. Also of interest to the *Construction* sector.
GIOPTIM	• Drawing up of an expert report on statistical sampling.
GSCUPM	• Development of a series of tasks within the project "Emergency service and operational support GEISER 2007". • GEISER Integration Phase 1 – Alarm functionality, automatic detection, BDAGEISER interface and other developments. • Studies of the hierarchy of physical and logical networks as complex systems in the optical telecommunications network. • Study of complex systems. Application to the technological domain. • Telnet team building and other development priorities.
InterTech	• Modelling of microstructure optical fibres.
mat+i	• Improvements in the THELSI electrolytic cell simulation package. Also of interest to the *Metal & Machinery* sector. • Scientific-technical assessment of databases (I-II). Also of interest to the *Management & Tourism* sector.
MODESI	• Obtaining symbolic solutions with Mathematica for the programming of electrical machines.

continued...

Psycotrip	• Analysis and design of an architectural framework for the publication of documents on the internet and their integration with external management systems. • Analysis of the requirements and technical design of the supporting architecture needed for the Barajas Airport Air Traffic Management Centre in Madrid. • Basic edition of documents via the internet and analysis of their possible application in template generation. • Development of a demonstration program about the software known as "Editor, driver and process viewer". • Development of MatesLab. • Development of TutorMates (I-II). • Evaluation report on the R&D activities of the Arsys Internet, S.L. company.
RUTYMETA	• Contract for consulting and technical support. Also of interest to the *Logistics & Transport* sector.
TAMI	• High resolution waferlevel reflowable EDoF camera module (WAFLE DONG I). • Interpolation and noise disposal algorithm of RAW digital imaging.

3.4.8.4
Training expertise

GOMA	• Applications of mathematical programming.
HYPCHAOP	• Signal and digital image processing through wavelet transformation. Also of interest to the *Biomedicine & Health* and *Metal & Machinery* sectors.
M2NICA	• i-MATH Intensive course on free software for Science and Engineering: Salome (I-III).
MAI	• Finite element methods: analysis, software and applications in Engineering (I-III). • i-MATH Intensive course on free software for Sciences and Engineering: Elmer (I-II).
mat+i	• i-MATH Intensive course on free software for Sciences and Engineering: Code_Aster (I-III).
MODESMAN	• Statistical data analysis with StatServer.
modestya	• i-MATH Intensive course on free software for Science and Engineering: R (I-III).
Psycotrip	• Editor, driver and process viewer training course.

3.4.9
Logistics & Transport

In terms of proven experience, technological transfer to the *Logistics & Transport* sector reaches a figure of 6%, but regarding supply the figure rises to 11%. According to our database, 31 direct contracts with industry and 5 training courses were developed by the research groups for this sector (reaching a total of 42 contracts and 15 training courses if the secondary sector *Logistics & Transport* is also included), many of them aimed at solving problems of transport route planning and optimization of production systems. The main applications of mathematical tools in *Logistics & Transport* are described below:

- optimization of resources;
- location of services;
- logistics and planning work;
- production planning;
- transportation planning and route optimization; vehicle traffic analysis; fleet optimization;
- simulation of rail traffic; location of railway sites;
- optimization and planning of loading and unloading;
- management and distribution of goods;
- shift allocation to employees;
- data protection systems;
- simulation of fluid-structure interaction problems;
- reliability and maintenance of railway infrastructure;
- technologies for urban transport.

3.4.9.1
Clients

- ACTIVIDAD PORTUARIA DE BARCELONA
- ADMINISTRADOR DE INFRAESTRUCTURAS FERROVIARIAS (ADIF)
- AMAYUELAS
- AMPO
- BINTER CANARIAS
- CÁMARA NAVARRA DE COMERCIO E INDUSTRIA
- CARNES ESTELLÉS
- CASTROSÚA
- COMPLEX SYSTEMS
- CONSULTORÍA Y COMUNICACIONES
- ESCUELA AULA
- EUSKO TRENBIDEAK
- EUSKOTREN
- FERROCARRILES DE VÍA ESTRECHA (FEVE)
- FUNDACIÓN DE LOS FERROCARRILES ESPAÑOLES
- INSTITÜT FUR TECHNO UND WIRSTCHAFT MATHEMATIK
- INSTITUTO DEL ENVASE, EMBALAJE Y TRANSPORTE (ITENE)
- MERCAZARAGOZA
- METRO BILBAO
- NORTAGRO
- RED NACIONAL DE FERROCARRILES ESPAÑOLES (RENFE)
- SCHNEIDER ELECTRIC

- SEMICROL
- SOCIEDAD PARA EL DESARROLLO REGIONAL DE CANTABRIA (SODERCAN)
- TRANSPORTES MUNICIPALES DE BARCELONA
- UNIVERSIDAD DE LA LAGUNA
- UNIVERSIDAD DE VIENA

3.4.9.2
Research groups

Groups with experience in the following section are:
ACEIA, CYOPT, DECYL, EOPT, GEUVA, GOMA, GPB97, GSO, mat+i, MCS-UAB, MODESI, MODESMAN, modestya, PROMALS, RUTYLO, RUTYMETA, TTM.

3.4.9.3
Consulting expertise

ACEIA	• Development of a railway simulation package for passenger transport route optimization. • Development of a tool that allows the analysis of the effect of the characteristics of rail network infrastructure and trains in the cost and income of providing a service and initial approximation to an intermodal comparison.
CYOPT	• Optimization of NORGRAFT PACKAGING production system (I-II).
DECYL	• Development of algorithms for optimal scheduling. • Logistic analysis of a pick-up service of used tires.
EOPT	• Allocation of work schedules to EuskoTren train drivers (I-II). • Design and development of an operational software tool for the annual optimal management of FEVE station shift workers. *Jointly with TTM*. • Design of a software tool for the annual planning of driver service. • Lists of annual tasks for bus and tram drivers. *Jointly with TTM*. • Lists of annual tasks for EuskoTren shift workers. *Jointly with TTM*.
GEUVA	• Collaboration and technical assistance in the conceptual, methodological and functional applications of the Asset Management Plan for ADIF. • Definition, validation and representation of indicators of reliability, maintainability and availability of rail infrastructure. • FIMALAC Project. Reliability, analysis and maintenance of the electrification system of RENFE under a RCM methodology (I-III).
GOMA	• Design and implementation of an automatic system to control the operational management of Binter Canarias. Also of interest to the *Metal & Machinery* sector.

continued…

GSO	• Allocation of shifts in industrial plants of Schneider Electric Spain S.A., Burlada and Puente la Reina. Also of interest to the *Energy & Environment* sector.
mat+i	• Ecotrans: Technologies for urban transport.
MCS-UAB	• Obtaining the optimal vehicle fleet for the distribution and delivery of hydrocarbon fuels in the Canary Islands. Also of interest to the *Energy & Environment* sector.
modestya	• Design and implementation of leadership, planning, monitoring and control techniques in order to achieve effective teams. Also of interest to the *Food* sector. • Value analysis applied to project management.
PROMALS	• AIMSUM: Definition of management policies for public transport in order to minimise fuel consumption. • AIMSUM: Design and implementation of school transport routes.
RUTYLO	• Automatic design of vehicle routing for the distribution of freight.
RUTYMETA	• Advanced software for the optimization of vehicle routing. • Study on the "Ciclo del Vagón" (time cycle of loading and unloading of railway freight cars).
TTM	• Annual planning for underground train drivers in Bilbao. • Annual work planning for train and tram drivers of EuskoTren. • Allocation of daily work schedules for EuskoTren train drivers. • Design and development of operational software tool for the annual optimal management of FEVE station shift workers. *Jointly with EOPT*. • Lists of annual tasks for bus and tram drivers. *Jointly with EOPT*. • List of annual tasks for EuskoTren shift workers. *Jointly with EOPT*.

3.4.9.4
Training expertise

GOMA	• Masters in logistics.
GPB97	• Cost allocation in optimization models.
MODESI	• Analysis and prediction of time series. • Masters in logistics: time series.
MODESMAN	• Analysis of statistical data with S-PLUS. Also of interest to the *Public Administration* sector.

3.4.10
Management & Tourism

Regarding *Management & Tourism*, 6% of supply and 8% of proven expertise of the research groups in the Consulting Platform correspond to this sector. In fact, the research groups have developed 45 contracts and 2 training courses related to *Management & Tourism* (reaching a total 62 contracts and 5 training courses if the secondary sector *Management & Tourism* is also included). Below is a list of some of the most widespread applications of mathematical technology in this sector:

- educational logical development;
- Economic Geography;
- characterisation of population habits;
- studies of professional integration;
- evolution of demographic rates; population projections;
- identification and planning of traffic in cities;
- simulation of service user flow;
- optimal task sequencing;
- calculation of house price indices; automated housing assessment;
- analysis of sport training parameters: performance in competition, referees visual acuity, reaction time, technologies for high performance athletes;
- design, development and analysis of surveys;
- statistical analysis of the behaviour of tourism networks;
- market volume studies;
- location of services;
- resources optimization.

3.4.10.1
Clients

- ALICER-ITC
- ASOCIACIÓN DE CLUBS DE BASKET
- CIRSA
- EQUANTIA GLOBAL BUSINESS
- GRAN CASINO DE BARCELONA
- IDOM
- INSTITUTO DE FORMACIÓN Y ESTUDIOS SOCIALES (IFES)
- INSTITUTO GALEGO DE ESTATÍSTICA (IGE)
- INSTITUTO NAVARRO DE SALUD LABORAL
- INSTITUTO VASCO DE ESTADÍSTICA (EU-STAT)
- JUNTA DE CASTILLA Y LEÓN
- KUKUXUMUSU
- NOVOTEC CONSULTORES
- SECRETARÍA XERAL DE EMIGRACIÓN (XUNTA DE GALICIA)
- TURGALICIA
- UNIVERSIDADE DE SANTIAGO DE COMPOSTELA
- VODAFONE ESPAÑA

3.4.10.2
Research groups

Groups with experience in the following section are:
DECYL, EE, GEUVA, MCS-UAB, MODESMAN, modestya, PROMALS.

3.4.10.3
Consulting expertise

DECYL	• Analysis of the impact of pregnancy on absenteeism. • Design, implementation and validation of an algorithm to quantify the risk of participating in the running of the bulls. • Quality study in Vodafone Spain. Consumer typology of customers. • Social and labour causes of sick leave reported in Navarra during the years 1998–2000. • Statistical modelling of industrial accidents in Navarra in the period 1989–1999.
EE	• External evaluation of the extrapolation methodology of the family budget survey and family expenditure statistics. Family budgets. Also of interest to the *Economics & Finance* and *Public Administration* sectors.
GEUVA	• General linear models to describe the evolution of demographic rates. Also of interest to the *Public Administration* sector. • State-space models to describe fertility rates in Castile and León. Also of interest to the *Public Administration* sector. • Stochastic models to obtain demographic projections in small areas in Castile and León. Also of interest to the *Public Administration* sector.
MCS-UAB	• ACB bets. • Report on the Shuffle Star automatic shuffler.
MODESMAN	• Supervision and technical assessment to ensure the relevance and improvement of consumer reports (IC) and competitive annual positioning (IPAC).
modestya	• Assessment in the design and obtaining of data in a survey on the population of the municipality of Arnoia (Ourense) into the attitude of citizens towards sustainable development in the municipality. Also of interest to the *Energy & Environment* and *Metal & Machinery* sectors. • Assessment of the design and analysis of survey data. • Determination of the sampling size in the analysis of the characterisation of lightweight packaging waste. Also of interest to the *Energy & Environment* sector. • Methodological design and analysis of survey results to enable quantification of the volume of tourism and hiking in Galicia in 2007, 2008 and 2009 (I-III). • Programs for training, multicultural integration for immigrants, and collaboration in statistical matters concerning migrations. • Research study into the market volume generated by hiking in Galicia in 2004.

continued...

	• Statistical study of hiking in Galicia – Hiking 2001, 2002, 2003, 2004 and 2005 (I-V).
	• Statistical study on summer tourism in Galicia – Destination 2003, 2004 and 2005 (I-III).
	• Statistical study on the volume of tourism and hiking in Galicia during 2006. Surveys: destination, origin and hiking.
	• Statistical study on tourism in Galicia – Destination 2000 and 2002 (I-II).
	• Statistical study on tourism in Galicia – Origin 2000, 2001, 2002, 2003, 2004 and 2005 (I-VI).
	• Statistics for tourist spending in overnight stays in hotels in Galicia in 2000, 2002 and 2004 (I-III). Also of interest to the *Public Administration* sector.
	• Survey design and analysis of data on citizens' attitudes towards sustainable development in Agenda 21 of the district of Pontevedra (councils of Barro, Pontecaldelas and Vilaboa).
	• Tourism statistics for overnight stays in hotels in Galicia during 2001, 2003 and 2005 (I-III).
PROMALS	• Development of research and computer programming for issuing instant electronic lottery tickets.

3.4.10.4
Training expertise

GEUVA	• Statistical analysis with computer application SPSS.
modestya	• SPSS 15.0.

3.4.11
Metal & Machinery

The *Metal & Machinery* sector represents 10% of the total supply from the groups and 14% in terms of proven experience, constituting the third most representative sector according to experience. Specifically, 62 contracts were developed and 22 training courses were conducted by research groups in this sector (reaching a total of 72 contracts and 25 training courses if the secondary sector *Metal & Machinery* is also included), covering a range of projects related to the field of materials, naval engineering, aeronautics and automotive engineering, many of them focused on CAD/CAE and numerical simulation techniques. Among the many applications of mathematical tools in the *Metal & Machinery* sector, we highlight some of the most important in the following list:

• mechanical or structural calculations; 3D structures; calculation of structures with reduced 1D and 2D models (beams, plates, plaques); contact, adhesion and friction between structures;

- elastic, viscoelastic and viscoplastic materials; behaviour laws; insulating materials;
- damage to structures: patterns of erosion and damage; mechanical calculations for the identification of cracks and fractures;
- heat transfer; thermal or thermodynamic calculations;
- Computational Fluid Dynamics (CFD); simulation of fluid-structure interaction problems;
- acoustic and vibro-acoustic calculations;
- simulation in electromagnetism; applications in electronic engineering, such as simulation of electronic devices;
- simulation of manufacturing processes: tamping, forging, etc.;
- numerical simulation of metals and ferroalloys casting;
- thermoelectric simulation of aluminium electrolysis cells;
- simulation of metallurgical electrodes;
- numerical simulation of induction furnaces;
- material purification, industrial grinding;
- active and passive control of noise; development of NVH systems for reduction of noise and vibration in vehicles;
- vehicle steering wheel design;
- numerical classification of burns from airbags;
- calculation of aerodynamic coefficient; numerical simulation of air flows around vehicles in paint booths;
- termohidrodinamic analysis of axial and radial bearings for ship propulsion systems;
- analysis of efficiency and reliability of ships;
- simulation and design of sails and kites;
- numerical characterisation of materials;
- numerical validation of ISO standards such as the overturning of buses;
- simulation of freezing and thawing processes;
- thermomechanical behaviour of heat exchangers; piezoelectric materials;
- design of air conditioning equipment in vehicles;
- production planning;
- control and optimization of products or processes;
- quality control; non-destructive inspection techniques;
- applications in port engineering.

3.4.11.1
Clients

- ADVANCED DYNAMICS
- AERTEC
- AIMEN
- ALCOA-INESPAL
- BALIÑO

- CANDEMAT
- CASTROSÚA
- CENTRE NATIONAL D'ÉTUDES SPATIALES (CNES)
- CIS FERROL

- CITRÖEN HISPANIA
- DALPHI METAL
- DESARROLLOS TÉCNICOS INDUSTRIALES DE GALICIA (DETEGASA)
- ESTRUCTURAS METÁLICAS CARLOS BAYO
- FERROATLÁNTICA
- FICOTRANSPAR
- FINSA
- FUNDACIÓN CESGA
- FUNDACIÓN CIDAUT
- IKERLAN
- INDUSTRIA DE DISEÑO TEXTIL (INDITEX)
- INSTITUTO TECNOLÓGICO DEL MUEBLE Y AFINES (AIDIMA)
- INTELLIGENT ADVISORS
- LABORATORIO OFICIAL DE METROLOGÍA DE GALICIA
- LUCENT TECHNOLOGIES DE MADRID
- MICRONICS
- NAVANTIA
- NAVARRA DE COMPONENTES ELECTRÓNICOS (NACESA)
- NOVOTEC CONSULTORES
- PAPELERA DEL BESAYA
- PERAMA INGENIERÍA SLU
- RIERA NADEU
- RUSSULA
- S.A.T. LINTEC
- SIDENOR
- SOCIEDAD PARA EL DESARROLLO RE-GIONAL DE CANTABRIA (SODERCAN)
- SY
- TECNOLOGÍAS AVANZADAS INSPIRALIA
- UNIVERSIDAD DE CÁDIZ
- UNIVERSIDADE DA CORUÑA

3.4.11.2
Research groups

Groups with experience in the following section are:
ACEIA, CAG, CYOPT, EE, Funaphy, GEUVA, KINETIC, M2NICA, M2S2M, MAI, mat+i, MCS-UAB, MODES, MODESI, MODESMAN, modestya, MOSISOLID, RUTYLO, RUTYMETA, TAMI, TTM.

3.4.11.3
Consulting expertise

ACEIA	• Accelerated time simulation package for the movement of passengers inside Málaga airport.
CAG	• New features for the industrial CAD/CAM/CAE environment for CSIS matrix generation: edge measurement, illumination, smoothing and offsetting. Also of interest to the *ICT* sector.
CYOPT	• Optimization of Besaya Paper Mill production system.
EE	• Technical report on the analysis of reclaimed parts.
Funaphy	• Acoustical study of the attenuation capacity of an acoustic screen formed of separate pipes made of absorbent materials and with varying diameter. Also of interest to the *Construction* sector.

continued...

GEUVA	• Report on the statistical validity of control tests.
KINETIC	• Mathematical model for the FICOSA windscreen de-icer. *Jointly with MCS-UAB.*
M2NICA	• Evaluation of a dynamic simulation model for bodies in rotation. • Numerical surface fitting algorithms for co-ordinate measuring machines.
MAI	• Conceptual design of a heat microexchanger by means of numerical simulation. • Dynamic analysis of heating ovens for rolling mills. • Numerical simulation of air flows around a vehicle in paint spray booths. • Preliminary analysis of the complete model for the heating of billets in rolling mill ovens. • Thermo-hydrodynamic analysis of axial and radial bearings for propulsion systems in ships.
mat+i	• Analysis of thermoelectric simulation of an electrolytic cell. • Automatic calculation of numerical definition of moulds for flexible materials with rigid grafts and complex surfaces. *Jointly with MOSISOLID.* • Calculation of tissue damage caused by exposure to hot gases. • Development and implementation of NVH systems for reducing noise and vibration in buses. • Elasto-acoustic numerical simulation with finite elements using parallel architectures. Application to active noise control. • Electromagnetic and fluid-thermal numerical simulation of processes in the silicon industry. • Mathematical model for continuous casting of ferro-alloys. • Mathematical modelling and computer-assisted simulation of electrodes, castings and purification processes in the Silicon industry (I–VI). • Mathematical modelling and computer-assisted simulation of processes in the silicon industry. • Mathematical modelling and computer-assisted simulation of the ELSA electrode and of silicon purification processes (I-II). • Numerical simulation of electrolytic cells and aluminium castings. • Technological-scientific consultancy regarding the development of new heating and refrigeration techniques for the formation of plastics using extruding technology. • Thermal and mechanical simulation of a heat exchanger. • Thermoelectric analysis of energy balance for different populations of cathodes in aluminium electrolysis cells. Setting future parameters. • Thermoelectric simulation of the Sabón electrode (I–IV).
MCS-UAB	• Mathematical model for the FICOSA windscreen de-icer. *Jointly with KINETIC.* • Study of the forces in a RINA Series 200 vertical axis centrifuge.
MODES	• Design of experiments in electrolysis series: application to the study of resistance modifiers in vats. • Technology and efficiency analysis of ships reliability.

continued...

MODESI	• Simulation of compound sections by means of "double T" profiles with similar mechanical features.
modestya	• Application of statistical methods to the optimization of hydraulic resources from the Xallas river (I-II). • Assessment of the methodology for determining the material composition and degree of humidity in recovered cardboard and paper. Also of interest to the *Energy & Environment* sector. • Design of analysis protocol of samples of children's clothing. • Statistical analysis for the monitoring and control of timber production processes. • Statistical analysis of the static and dynamic variables related to the electrical operation and casting systems in ferro-alloy furnaces (I-III). • Statistical analysis of variables related to the operation of ferro-alloy furnaces (I–II). • Statistical assessment for quality control of material recovered from lightweight packaging. Also of interest to the *Energy & Environment* sector. • Statistical techniques for the control and modelling of the static and dynamic variables associated with electrical operation and casting systems in ferro-alloy furnaces (I-IV). Also of interest to the *Energy & Environment* sector.
MOSISOLID	• Automatic calculation of numerical definition of moulds for flexible materials with rigid grafts and complex surfaces. *Jointly with mat+i.*
RUTYLO	• Long-term analysis of workshop problems.
TAMI	• Adhesion correction for adaptive windows. • On board optimization/compression and restoration on land and in stereo at low b/h (I-II).
TTM	• Application of OES-PDA technology in the evaluation of bearing steel homogeneity.

3.4.11.4
Training expertise

GEUVA	• Statistical techniques applied to sample design. Multivariate analysis and model retrieval. Risk-benefit analysis. Cost-benefit analysis (I-II). Also of interest to the *Logistics & Transport* sector. • Statistical tools with SPSS. Also of interest to the *Logistics & Transport* sector.
M2NICA	• i-MATH Intensive course on free software for Science and Engineering: Python programming oriented to engineering (I-III). Also of interest to the *Biomedicine & Health, Economics & Finance, Energy & Environment, Food, ICT, Logistics & Transport* and *Public Administration* sectors.

continued …

	• Scientific visualisation with VTK. Also of interest to the *Biomedicine & Health, Energy & Environment, ICT* and *Logistics & Transport* sectors.
M2S2M	• Introduction to FreeFEM ++ software. Also of interest to the *Construction, Energy & Environment, Food* and *Logistics & Transport* sectors.
mat+i	• Course on management of simulation applications for casting C2D and C3D. • Course on Nanotechnology & Mathematics. • Finite element method. • i-MATH Short course on numerical simulation in electromagnetism and industrial applications (I-II). • Management of THELSI 3D software. • Mathematical models in acoustics.
MODES	• Applied basic statistics. Also of interest to the *Energy & Environment* sector. • Descriptive and exploratory data analysis, parametric and non-parametric statistical inference. • Experimental design and analysis. Also of interest to the *Energy & Environment* sector. • Statistical analysis applicable to experiments. Also of interest to the *Energy & Environment* sector. • Training course on statistical support tools for the functionality of technological surveillance.
MODESMAN	• S-PLUS and its statistical applications.
RUTYMETA	• Data analysis and design of experiments.

3.4.12
Public Administration

The *Public Administration* sector is the most relevant sector in terms of proven experience in knowledge transfer, accounting for 26% of total activities (contracts and training) carried out by groups in all the previously considered sectors. The groups' expertise in knowledge transfer has been put into practice through the establishment of 85 direct contracts and 66 training courses for industry (reaching a total of 98 contracts and 70 training courses if the secondary sector *Public Administration* is also included).

The intensive collaboration within this sector is mainly due to its tradition in using statistical techniques for advising governmental organisations and public hospitals in the development and design of surveys, social studies and general data processing. The main applications developed by the research groups in this area are the following:

- design, development and analysis of surveys;
- macro surveys; electronic voting techniques (e-voting);
- application and use of databases; data mining;
- statistical data confidentiality; electronic security;
- development of educational logics;
- bibliometric analysis;
- vehicle traffic planning in cities; sustainable mobility;
- statistical models of occupational injury; estimation of unemployment rates; calculation of population projections;
- location of services;
- optimal sequencing of tasks;
- sustainability in division of land; geographic information systems;
- Economic Geography; house price indexes; automatic assessment of housing; durability of buildings;
- analysis of parameters related to sport practice: performance in competition, referees visual acuity, reaction time, technologies for high performance athletes.

3.4.12.1
Clients

- AJUNTAMENT DE L'HOSPITALET DE LLOBREGAT
- AULA DE INFORMÁTICA DEL CAMPUS DE LUGO – UNIVERSIDADE DE SANTIAGO DE COMPOSTELA
- AXENCIA PARA A CALIDADE DO SISTEMA UNIVERSITARIO DE GALICIA (ACSUG)
- AYUNTAMIENTO DE NIGRÁN
- AYUNTAMIENTO DE OURENSE
- CÁMARA DE CUENTAS DE ANDALUCÍA
- CBS
- CENTRO DE INVESTIGACIONES MARINAS (CSIC)
- CENTRO REGIONAL DE ESTADÍSTICA
- CONSEJERÍA DE OBRAS PÚBLICAS (GOBIERNO DE ARAGÓN)
- CONSEJO SUPERIOR DE DEPORTES
- CONSELL SOCIAL (UdG)
- CONSELLERÍA DE EDUCACIÓN E ORDENACIÓN UNIVERSITARIA (XUNTA DE GALICIA)
- CONSELLERÍA DE SANIDADE (XUNTA DE GALICIA)
- CONSELLO SOCIAL
- CONSORCI URBANÍSTIC DEL CENTRE DIRECCIONAL DE CERDANYOLA
- DIPUTACIÓN DE LA CORUÑA
- DIPUTACIÓN DE LUGO
- DIRECCIÓN XERAL DE SAÚDE PÚBLICA (XUNTA DE GALICIA)
- EMPLEO DE CASTILLA Y LEÓN (ECYL – JUNTA DE CASTILLA Y LEÓN)
- ERYBA
- ESCOLA GALEGA DE ADMINISTRACIÓN SANITARIA
- ESCUELA DE ADMINISTRACIÓN PÚBLICA DE CASTILLA Y LEÓN (ECLAP – JUNTA DE CASTILLA Y LEÓN)
- EUROPEAN RAILWAYS
- EUROSTAT/STATISTICS NETHERLANDS
- FEB (ÁVILA, BILBAO, MADRID, MÁLAGA, MENORCA, VALLADOLID)
- FEMXA
- FUNDACIÓN CESGA
- FUNDACIÓN PARA A ORIENTACIÓN PROFESIONAL, A INVESTIGACIÓN E O DESENVOLVEMENTO TECNOLÓXICO, O EMPREGO E FORMACIÓN EN GALIZA (FORGA)

- GERENCIA DE SALUD DE LA JUNTA DE CASTILLA Y LEÓN
- GERENCIA DE SERVICIOS SOCIALES DE LA JUNTA DE CASTILLA Y LEÓN
- INSTITUT ESTADÍSTICA DE CATALUNYA (IDESCAT)
- INSTITUTO ARAGONÉS DE ADMINISTRACIÓN PÚBLICA
- INSTITUTO ARAGONÉS DE ESTADÍSTICA (IAEST)
- INSTITUTO CANARIO DE ESTADÍSTICA (ISTAC)
- INSTITUTO CÁNTABRO DE ESTADÍSTICA
- INSTITUTO DE ESTADÍSTICA DE ANDALUCÍA
- INSTITUTO DE ESTADÍSTICA DE NAVARRA (IEN)
- INSTITUTO DE LA PEQUEÑA Y MEDIANA IN- DUSTRIA DE LA GENERALITAT VALENCIANA (IMPIVA)
- INSTITUTO GALEGO DE ESTADÍSTICA (IGE)
- INSTITUTO NACIONAL DE ESTADÍSTICA (INE)
- INSTITUTO TÉCNICO AGRARIO (ITA)
- INSTITUTO VASCO DE ESTADÍSTICA (EUSTAT)
- JUNTA DE ANDALUCÍA

- JUNTA DE CASTILLA Y LEÓN
- LKS
- MINISTERIO DE VIVIENDA
- ONS
- SERVICIO ANDALUZ DE SALUD
- SOCIEDADE GALEGA PARA A PROMOCIÓN DA ESTATÍSTICA E DA INVESTIGACIÓN OPERA- TIVA (SGAPEIO)
- SOCIEDADE PARA O DESENVOLVEMENTO CO- MARCAL DE GALICIA
- STATGRAPHICS CONSULTING
- THALES
- TRABAJOS CATASTRALES
- UNIVERSIDAD AUTÓNOMA DE MADRID
- UNIVERSIDAD CARLOS III DE MADRID
- UNIVERSIDAD DE VALENCIA
- UNIVERSIDAD DE ZARAGOZA
- UNIVERSIDAD DEL PAÍS VASCO – EUSKAL HERRIKO UNIBERTSITATEA
- UNIVERSIDAD POLITÉCNICA DE CATALUNYA
- UNIVERSIDADE DE SANTIAGO DE COM- POSTELA (USC)
- UNIVERSIDADE DE VIGO

3.4.12.2
Research groups

Groups with experience in the following section are:
CAG, CODA, DEPREN, EE, EOPT, GEUVA, GIOPTIM, GNOM, GOMA, GOR, GPB97, GSO, INFERES, mat+i, MCS-UAB, MODES, MODESI, modestya, MO- SISOLID.

3.4.12.3
Consulting expertise

CAG	• Determination and planning of mobility in the city of Vitoria by means of EMME/2. Also of interest to the *Logistics & Transport* sector.
CODA	• Compositional data package (CoDaPack).
DEPREN	• Exclusive training for all teams and individuals in Spain who acquire the "Dar- trainer Pro" Program. Also of interest to the *ICT* sector.

continued…

EE	<ul><li>Analysis of the Canadian methodology of estimators using longitudinal information.</li><li>Demographic methods. Compound estimators and SPREE estimators. Also of interest to the *Management & Tourism* sector.</li><li>Development of computer programs for estimation based on longitudinal models and small area estimation.</li><li>Development of informational statistical programs for obtaining estimators at a county level for several sets of statistical data to be carried out by EUSTAT.</li><li>Estimation of cultivated land and crop yields in Navarra by area sampling with irregular segments and remote sensing.</li><li>Estimation of the number of unemployed, employed and inactive people in Navarra according to county.</li><li>Evaluation methodology for estimators by simulation.</li><li>Modelling housing availability with small areas (I-II). Also of interest to the *Management & Tourism* sector.</li><li>PRA longitudinal statistical methodology.</li><li>Probabilistic methods of record data linkage.</li><li>Small area estimation in Navarra (I-II).</li><li>Small area sampling.</li><li>Small area study in the technological innovation survey and development of probabilistic methods of record data linkage.</li><li>Statistical modelling for the valuation of collective housing in Álava. Also of interest to the *Management & Tourism* sector.</li><li>Technical and methodological assistance in the estimation and implementation of statistical indicators for small geographic areas of EUSTAT operations.</li><li>Technical assessment in statistics and data analysis.</li><li>Technical assistance for obtaining small area estimators.</li><li>Technical assistance for obtaining small area estimators. Application of small area estimation in the statistical operation of the PRA. Also of interest to the *Management & Tourism* sector.</li><li>Traditional direct and indirect estimators.</li></ul>
EOPT	<ul><li>Analysis of automatic imputation in the IPI.</li><li>Business plan and market study for an institutional gift shop.</li><li>Imputation of missing data in the R&D survey.</li><li>Imputation of time series with application to various statistical indices.</li><li>Review of the operation of a survey of tourist establishments of the CAV.</li></ul>
GEUVA	<ul><li>Statistical analysis of sub-provincial socioeconomic indicators in Castile and León. Assessment of its use to improve estimates of unemployment rates obtained from the labour force survey.</li></ul>
GIOPTIM	<ul><li>Statistical assessment and drawing up of a sampling manual for the IMPIVA internal audit.</li><li>Statistical sampling in IMPIVA (I-III).</li><li>The drawing up of a manual on statistical sampling and statistical assessment at the Audit Chamber in its activities during 2005.</li></ul>

continued…

GNOM	<ul><li>Consultation and technical assistance in statistical confidentiality of European aggregates for structural business statistics.</li><li>Consultation and technical assistance in statistical confidentiality of tabular data.</li><li>Extension of a CTA code (Controlled Tabular Adjustment) for statistical confidentiality of European aggregates for animal production statistics.</li><li>Implementation of a tool for statistical confidentiality of European aggregates for structural business statistics and balance of payment.</li><li>Safe configuration of tabular data.</li></ul>
GOMA	<ul><li>Anonymisation group.</li><li>Data protection.</li><li>Rounding off tables.</li><li>Survey purification.</li></ul>
GOR	<ul><li>Research into techniques and automatic processes for statistical secrets in a multi-dimensional database environment and dynamic dissemination of information.</li></ul>
GSO	<ul><li>Functionality development of the computer application PROASCYL: calculation of district medical emergencies in Castile and León. Also of interest to the *Biomedicine & Health* sector.</li><li>Functionality development of the computer application PROASCYL: Calculation of the medical districts of Castile and León. Also of interest to the *Biomedicine & Health* sector.</li><li>Modelling and heuristic problem solving for large-scale territorial design.</li><li>Optimization models for allocation of shifts of health staff. Also of interest to the *Biomedicine & Health* sector.</li><li>Optimizing holiday leave cover in the management of shifts in care homes. Also of interest to the *Biomedicine & Health* sector.</li><li>Scheduling of duty staff for primary care teams and general medical staff. Also of interest to the *Biomedicine & Health* sector.</li><li>Study of different location models of regional health services in Castile and León and implementation of software (PROASCYL). Also of interest to the *Biomedicine & Health* sector.</li></ul>
INFERES	<ul><li>Study of the impact of the gap in digital technology in the city of Ourense.</li><li>Survey of customer satisfaction with solid urban waste collection in the council of Nigrán.</li></ul>
MCS-UAB	<ul><li>Guidelines for the fair apportioning of compensation for compulsory purchase of assets.</li><li>Profile of women who are associate members of L'Hospitalet.</li></ul>
MODES	<ul><li>Analysis and study of the problems detected in IGE responsibility sampling surveys.</li><li>Methods of variance estimation in IGE managed surveys.</li></ul>

continued…

MODESI	<ul><li>Fair drawing of lots for allocation of social housing.</li><li>Statistical information system on population and social affairs. Statistical atlas of the Pyrenees.</li></ul>
modestya	<ul><li>Analysis of labour market insertion of Galician University System graduates in 2003–2005 and 2005–2006 (I-II).</li><li>Collaboration agreement between the Galician Institute of Statistics and the University of Santiago de Compostela for a research project in small area sampling, with applications.</li><li>Division of contents associated with the training modules of new professional certificates.</li><li>Drawing up of didactic materials and guides for teachers and students: assembly and installation of photovoltaic sites.</li><li>Improvement in the estimation and prediction of variables and parameters of interest in surveys conducted by the IGE. Small area sampling (I-II).</li><li>Incorporation of graduates within the labour market. Vision from company view point. Also of interest to the *Management & Tourism* sector.</li><li>Quality index of CESGA scientific production.</li><li>Inter-sector study on educational credits ratified for professional qualifications in the electrical and electronic installation and maintenance sector.</li><li>Interviews for the Galician Institute of Statistics. Also of interest to the *Management & Tourism* sector.</li><li>Labour market insertion of Galician University System graduates in 1996–2001. Also of interest to the *Management & Tourism* sector.</li><li>Methodological study of the correspondence between competitive units and jobs in the professional sector of mechanical manufacturing.</li><li>Reengineering of route allocation for network timetables for European railways. Also of interest to the *Logistics & Transport* sector.</li><li>Labour profile of Galician University System graduates according to qualification. Also of interest to the *Management & Tourism* sector.</li><li>Research and development of small area estimation methods.</li><li>Statistical methodology in the different phases of PanIL (2007–2011). Also of interest to the *Management & Tourism* sector.</li><li>Study and valuation of the accident rate involving wild animals in the provincial road network of Lugo in relation to use and fragmentation of habitat.</li><li>Study by sector of the division of contents associated with the training modules of new professional certificates.</li><li>Study of labour market insertion of Galician University System graduates 2006–2007. Adaptation of the methodology to the Galician Statistical Plan 2007–2011. Also of interest to the *Management & Tourism* sector.</li></ul>
	<ul><li>Study of labour market insertion of Galician University System graduates 2007–2008. Also of interest to the *Management & Tourism* sector.</li><li>Study of the impact of work experience on the insertion of USC graduates into the labour market. Also of interest to the *Management & Tourism* sector.</li><li>Study on renewable energy by sector.</li></ul>

continued...

MOSISOLID	• Modification of the formula for the evaluation criteria of economic tenders and, in general, of abnormal or disproportionate values. Also of interest to the *Economics & Finance* sector. • Study for the valuation of economic proposals. Also of interest to the *Economics & Finance* sector.

3.4.12.4
Training expertise

DEPREN	• 3rd International preparation course on high-level physical training for basketball. Also of interest to the *ICT* sector. • 3rd International specialisation course on high-level physical training for basketball. Also of interest to the *ICT* sector. • Advanced basketball coaching course. Also of interest to the *ICT* sector. • Advanced coaching course (I–II). Also of interest to the *ICT* sector. • Course on physical training for basketball. Also of interest to the *ICT* sector. • Official ratification course on first level basketball coaching. Also of interest to the *ICT* sector. • Specialisation courses. Ongoing coaching course (I–III). Also of interest to the *ICT* sector. • Training course for trainers. Also of interest to the *ICT* sector.
EE	• Statistics with R.
EOPT	• Multivariate analysis. Factorial and classification techniques.
GEUVA	• Advanced methodology: time series and logistic regression. • Introduction to Statistics. • Linear methods: regression and classification. • Statgraphics in control and statistical analysis of analytical data. Also of interest to the *Biomedicine & Health* sector. • Statistical analysis with SPSS (I–VI). • Statistical techniques applied to labour mediation. • Statistical tools with SPSS. Also of interest to the *Management & Tourism* sector. • Using statistics to help in labour mediation. Also of interest to the *Management & Tourism* sector.
GIOPTIM	• Data mining. • Introduction to R. • Knowledge extraction techniques in large databases. • R environment. Introduction to R-Commander. • Study and dissemination of new research and analysis instruments to be applied in the 2007–2010 operations of the Andalusian Statistical Plan. • Training course in sampling.

continued...

GPB97	<ul><li>Executive management training course.</li><li>Statistics in secondary education.</li></ul>
mat+i	<ul><li>Numerical methods for hyperbolic equations. Theory and applications.</li></ul>
MODES	<ul><li>Integration course on the upper scale of statistics: SPSS Module.</li></ul>
MODESI	<ul><li>Introduction to basic Statistics.</li></ul>
modestya	<ul><li>An introduction to bootstrap.</li><li>Analysis of functional data (I–II).</li><li>Applied statistical sampling.</li><li>Basic biostatistics (I–II).</li><li>Basic course on small area estimation.</li><li>Course on statistics with SPSS.</li><li>Inferential statistics.</li><li>Initiation in SPSS with Windows (I–II).</li><li>Introduction to free software. Use of R statistical system.</li><li>Introduction to the use and programming of the R statistical system.</li><li>Multivariate analysis.</li><li>R environment.</li><li>Regression methods (I–IV).</li><li>Seminar on Statistics and Probability for secondary school teachers.</li><li>Statistical analysis with R (I–III).</li><li>Statistics for secondary school mathematics teachers.</li><li>Statistics for secondary school teachers – Use of MS EXCEL statistics (I-II).</li><li>Statistics with functional data.</li><li>Statistics with SPSS (I–II).</li></ul>

3.5
Expertise in software development

Most of the methods and mathematical techniques applied to different sectors of economic activity shown in the previous section lead to the development of customised software packages for solving the existing challenges and needs in companies.

The use of custom-made software is normally required by companies, since it simplifies the use of solutions developed based on advanced mathematical techniques. Furthermore, it includes user-friendly interfaces to facilitate the use of technical solutions for a wide range of skill bases and end-users and, in many cases, a training course for the company staff required to work with the customised package. Therefore, the development of specialised software is an essential part of the work of research groups engaged in knowledge transfer.

This section contains a description of proprietary and free software developed by the research groups that took part in the survey. Among the 111 software packages developed by the research groups in the Consulting Platform, 39 of them were transferred to industry. Therefore, as in the previous section, a list of clients is presented below. The research groups with proven expertise in the development of customised software are shown, as well as a complete list of software developed by them. The packages are categorised according to economic sector in which they are mainly focused, while for the most general software, a "transversal sector" is included at the end of the section.

3.5.1
Clients

- AEROPUERTOS ESPAÑOLES Y NAVEGACIÓN AÉREA (AENA)
- ALCOA-INESPAL
- CANDEMAT
- CARNES ESTELLÉS
- CENTRO DE PESQUISAS DE ENERGIA ELÉTRICA (CEPEL-BRASIL)
- CENTRO NACIONAL DE ENERGÍAS RENOVABLES (CENER-CIEMAT)
- CIRSA
- COMPLEX SYSTEMS
- CONSEJERÍA DE EDUCACIÓN DEL GOBIERNO DE CANARIAS
- DALPHI METAL
- ENDESA
- ENS CACHAN
- EUROSTAT
- FERROATLÁNTICA
- FUJITSU
- FUNDACIÓN CESGA
- GEP-COMPAGNIE BANCAIRE
- GOBIERNO DE ARAGÓN
- GRANITOS MONTE FARO
- INDUSTRIAS GONZÁLEZ
- INSTITUTO CANARIO DE ESTADÍSTICA (ISTAC)
- INSTITUTO NACIONAL DE ESTADÍSTICA (INE)
- INVERNESS MEDICAL
- ITENE
- JUNTA DE ANDALUCÍA
- KUKUXUMUSU
- NEXANS IBÉRICA
- NORGRAFT PACKAGING
- NTNU
- ONS
- OPTIMIZATION TECHNOLOGIES
- PAPELERA DEL BESAYA
- PRODUCT SENSORY CONSULTING GROUP
- SALUD CASTILLA Y LEÓN (SACYL)
- SOCIEDAD PARA EL DESARROLLO DE CANTABRIA (SODERCAN)
- TRANSPORT SIMULATION SYSTEMS
- TRANSPORTES INTERURBANOS DE TENERIFE (TITSA)
- UNIVERSIDAD DEL PAÍS VASCO – EUSKAL HERRIKO UNIBERTSITATEA
- UNIVERSIDAD REY JUAN CARLOS

3.5.2
Research groups

Groups with experience in the following section are:
ACEIA, ADF, CAG, CODA, CYOPT, DDA, DECYL, EDANYA, EDnL, EE, EOPT, GMFN, GNOM, GOMA, GPB97, GRASS, GRID[ECMB], GSC, GSO, HYPCHAOP, INFERES, InterTech, KINETIC, M2NICA, M2S2M, MAI, mat+i,

MCS-UAB, MODES, MODESI, MODESMAN, modestya, MOSISOLID, OED-group, PROMALS, Psycotrip, RUTYLO, RUTYMETA, TAMI, TAPO, TTM, varidis.

3.5.3
Software development expertise

Biomedicine & Health	
GPB97	• Analysis of biological, biomedical and epidemiological data.
GRASS	• **AUCRF (package)**. Genomic profiling with Random Forest. • **bwsurvival (R package)**. Package designed for those situations where there are two events of interest, E1 and E2, and the objective centres on estimating the time of survival function between T2 and E2 based on time T1 to E1. • **dcens (R package)**. Package which estimates the survival function in a double blind data scheme. • **MB-MDR (package)**. Analysis of genetic interactions.
GRID[ECMB]	• **Epilinux**. Freely available operating system aimed especially at the use of epidemiological and biostatistical analysis tools. • **p3state.msm**. Inference in multi-state models in the field of medicine, that provides a greater understanding of the development and evolution of an illness over time. • **ROCRegression (R package)**. Package which determines the evaluation of covariates effect on the discrimination capacity of a continuous marker, measured by the ROC curve.
GSO	• **PROASCYL – Programa de Optimización de la Asistencia Sanitaria de Castilla y León**. Software which aims to assist in the optimal location of health services in Castile and León.
MCS-UAB	• Levenberg-Marquardt method for the nonlinear fitting of data, used as part of the ELISA spectrophotometer software for a company that manufactures instruments of diagnosis. • Optimization of the administration of medicines.
modestya	• **PMICALC – Post Mortem Interval Calculator**. Additive models and support vector machines (SVM) for calculating the post mortem interval.
MOSISOLID	• Resolution of bone formation models using finite elements.
OEDgroup	• **Biokmod (Mathematica application)**. Mathematical modelling in internal dosage, nuclear medicine and pharmacology.

continued...

TAPO	• Corneal topography reconstruction using altimetry or curvature data.

Construction

GMFN	• **FOLDMODELER and FOLDPROFILER (Mathematica and MATLAB applications)**. Modelling tools for natural folding.
HYPCHAOP	• **JOINLANDS**. Software for sustainable land subdivision: a list of land plot groupings is obtained from the land register, such that each group produces an optimal surface area according to certain restrictions.
mat+i	• **OPTICORTE**. Software for the optimization of cutting sections from granite blocks.
RUTYLO	• Software for concrete transport optimization: production, planning and supply from production plants to construction sites.

Economics & Finance

EOPT	• **BFC – Branch and Fix Coordination**. Resolution of very large dimension 0–1 mixed optimization models. • **CBD – Cluster Benders Decomposition**. Resolution of large dimension linear optimization models. • **CLD – Cluster Lagrangean Decomposition**. Determination of good benchmarks in large dimension 0–1 mixed optimization models.
GPB97	• Business simulators.
M2NICA	• **CUDA packages for option pricing by Monte Carlo methods**. Efficient implementation in GPUs of Monte Carlo methods for option pricing. • **CUDA packages for simulated annealing**. Implementation of adaptive simulated annealing type methods for calibration of stochastic volatility SABR type models. • Software for financial derivatives valuation such as options and bonds, based on Black-Scholes models and quantitative financial techniques.
MODES	• **AGLOSS**. Modelling of aggregate actuarial risk. Estimation of future loss due to accidents, and analysis of the viability of different insurance strategies for companies. • Software for analysing credit risk in the financial sector by means of the estimation of the default probability using survival analysis. • Software for analysing credit risk in the financial sector which, from a credit database, calculates the associated scoring model, on the basis for which new clients are classified.
MODESMAN	• **SPPA – Spatial Point Pattern Analysis**.

continued…

Energy & Environment	
CYOPT	• **SINEPAC.ERP (module)**. Software for optimizing a packaging production system in the paper and cardboard industry. • Software for production system optimization in a paper mill.
DDA	• **SOL**. Software for obtaining solar radiation maps on irregular orography by means of adapted triangulations. • **Wind3DCode**. Tri-dimensional simulation of wind farms for the drawing up of wind maps, and therefore the evaluation of potential wind energy production.
DECYL	• **WindHyGen**. Software for the economical assessment of hybrid energy systems composed of a wind farm and a system for the transformation, storage and recovery of energy using hydrogen technology.
EDANYA	• **HySea (web platform)**. Simulation via the internet of geophysical flows, such as floods, oil spills, spread of discharges, sediment transport, turbid currents, tsunamis, etc.
GMFN	• **GEOMAT2D (MATLAB application)**. Geostatistical modelling for bi-dimensional regionalised variables with several functionalities: statistical analysis, structural analysis and different kriging techniques. • **MTCLAB (MATLAB application)**. Quality analysis for tomographic data and inference of initial models in 2D tomographic experiments in transmission. • **VESLAB (MATLAB application)**. Deterministic and probabilistic resolution of the inverse problem in vertical electric soundings.
GSC	• Pollutant dispersion simulator: DBO-OD concentrations, concentration of a specific pollutant, etc. • Shallow water movement simulator. Calculation of averaged velocities and water column heights. • Two and three-dimensional thermo-hydrodynamic simulation for the resolution of two and three-dimensional Navier-Stokes equations with coupled thermal systems.
M2NICA	• **GLANUSIT Toolbox**. Simulation of the thermo-elasto-hydrodynamic behaviour of large masses of ice, such as glaciers and polar ice caps, based on shallow ice-type models. *Jointly with MAI.*
M2S2M	• **DAMFLOW**. Simulation for hydrodynamic flows. • **FREEFEM++ MEDIOAMBIENTAL**. Numerical solution of 3D environmental hydrodynamic flows.

continued…

MAI	<ul><li>**Comsol (Applications)**. Thermal simulations for the heating of the ground due to forest fires.</li><li>**GLANUSIT Toolbox.** Simulation of the thermo-elasto-hydrodynamic behaviour of large masses of ice, such as glaciers and polar ice caps, based on shallow ice-type models. *Jointly with M2NICA.*</li></ul>
mat+i	<ul><li>**AcousFEM**. Numerical solution of problems in acoustics and elastoacustics, including dissipative media.</li><li>**SC3D**. Three-dimensional simulation of a pulverized carbon boiler, including movement of gases, heat transfer by conduction, convection and radiation, and carbon combustion.</li><li>**THESIF**. Thermo-electromagnetic-hydrodynamic modelling of an induction oven.</li></ul>
MODES	<ul><li>**GEOEST**. Software for the analysis of spatial and spatial-temporal data: exploratory analysis, structural analysis and kriging prediction.</li></ul>
modestya	<ul><li>Real time environmental prediction of source of pollutant emissions using historic company database.</li></ul>
Food	
MCS-UAB	<ul><li>Software for the experimental design of sensorial analysis developed for a company involved in market studies.</li></ul>
MODESI	<ul><li>**CRDO**. Software for managing consignments of wine carrying protected designation of origin status.</li></ul>
ICT	
ACEIA	<ul><li>Connection of dynamic geometry systems and symbolic computation.</li><li>Demonstration and automatic discovery in Geometry.</li></ul>
CODA	<ul><li>**CoDaPack – Compositional Data Package**. Statistical analysis of compositional data applicable to areas where compositional-type multivariate data are used.</li></ul>
CYOPT	<ul><li>**PLCBAS**. Quadratic programming software that was subsequently included in the SCILAB package.</li></ul>
GNOM	<ul><li>Development of backup software in Fortran, C and C++.</li></ul>
Psycotrip	<ul><li>Development of symbolic software and information systems (databases).</li></ul>
RUTYMETA	<ul><li>**OptQuest**. General purpose optimization software based on heuristic algorithms.</li></ul>

continued…

Logistics & Transport	
ACEIA	• Analysis of the evolution of transport network connectivity. • Decision-making in railway signal boxes. • Simulation of passenger movements. • Simulation of railway traffic.
GOMA	• Software for school transport optimization which combines algorithms to find optimal routes and geographic information systems. • Software for work schedules optimization for public transport bus drivers conforming to both union conditions and company requirements.
MCS-UAB	• Software for managing data in a large-scale optimization problem using the Nelder-Mead method.
PROMALS	• **AIMSUM**. Urban traffic simulator that includes different optimization algorithms for the design of vehicle routes. • System to help decision-making aimed at the resolution of different types of vehicle itinerary problems and location of facilities.
RUTYLO	• **RutaRep**. Software to assist in the planning and design of distribution and delivery routes.
Management & Tourism	
DECYL	• **Encierrómetro**. Software to assess the risk of participating in the running of the bulls in Pamplona.
PROMALS	• Research development and computer programming for the issuing of instantaneous electronic lottery tickets.
Metal & Machinery	
ACEIA	• Inference engines and expert systems verification using algebraic techniques.
ADF	• **CFD 3D RANS**. Collaboration in the development of the DLR TAU code. • CFD software, application of numerical tools for the analysis and design of aerodynamic configurations.
CAG	• **CSIS (modules)**. Maintainance and support of CSIS CAD/CAM software: illumination module; conventional measurement module; cut measurement module; and reading system for the acquisition of data.
CYOPT	• Software for planning and optimization of wire coating phase with applications in the cabling industry.

continued...

HYPCHAOP	• **FAILDET**. Software for detection of partial and total breakage in rotor bars of industrial induction machines, analysing the signal associated to around 2 seconds of the startup of the machine.
InterTech	• Software for the analysis of the electro-magnetic field in nonlinear photonic devices.
KINETIC	• Semiconductor devices: software applicable to design and development of devices in electronic engineering sectors.
M2NICA	• Simulation of lubricated devices that form part of machines, such as axles or bearings, based on Reynolds-type models, also including cavitations and/or surface deformation.
MAI	• **ELMER (applications)**. Thermo-hydrodynamic simulation of heat micro-exchangers. Thermal simulations of heating ovens for steel billets. • Software for simulating mechanical and thermo-hydrodynamic behaviour of engine bearings solving partial differential equations corresponding to finite element methods, finite volume methods, and boundary element methods.
mat+i	• **AcousFEM**. Numerical solution to problems in acoustics and elasto-acoustics, including dissipative media. • **BREAB**. Software which simulates evolution of skin temperature and tissue damage, and classifies burns caused by the explosion of an airbag. • **C2D and C3D**. Packages for bi- and tri-dimensional thermo-mechanical simulation of aluminium casting behaviour. • **CEM2D**. Software for magnetic-hydrodynamic simulation of aluminium casting. • **COLADA**. Software which numerically simulates the thermal behaviour of a ferro-alloy casting. • **ELSATE**. Software for determining the distribution of current, temperature and mechanical forces in a radial section of an electrode. • **ELVA. Estabilidad Lateral al Vuelco de Autobuses**. Calculation of lateral stability against an overturning bus in both dynamic and static tests. • **MaxFEM**. Troubleshooting electromagnetism. • **PAMM. Propagación Acústica en un Medio Multicapa**. Software which enables the prediction of the acoustic behaviour of stratified media formed by materials of different characteristics. • **SON**. Parallel software for vibration calculation applied to active noise control. • **THELSI3D**. Software which enables the thermo-electric behaviour of an aluminium electrolysis reduction pot to be numerically simulated three-dimensionally.
TTM	• Finite elements software for solving direct and inverse problems in electromagnetism, acoustics, elasticity and fluids.

continued...

Public Administration	
EE	• Development of SAS and R macros for the estimation of variables by county, such as mortality rates, cultivated surface areas, industrial variables, valuation models and house pricing.
GOMA	• Data protection tool for European statistics institutes which determines the secondary withholding of data to ensure the protection of sensitive information when tables are published. • **TEIDE**. Purification of statistical surveys to guarantee that microdata subjected to statistical analysis are free of errors.
GPB97	• Management system by means of indicators.
TAMI	• **IPOL**. Online software for image processing. • **Megawave**. Public software for image processing.
Transversal sector	
DDA	• **SUS Code**. Smoothing and untangling of tetrahedral meshes.
EDnL	• **Green's functions**. Implementation of Green's functions to solve boundary problems for differential equations.
GOMA	• Rounding algorithm with linear relationships based on mathematical programming techniques.
GSC	• Objective function optimization using methods with and without derivates.
INFERES	• Software including statistical methodology, for example, spatial statistics, survival analysis, environmental statistics, econometrics, biostatistics, finance, routing problems, etc.
modestya	• **alpha-hull**. Generalisation of a convex envelope for a sample of points in a plane. • **DTDA**. Package that implements different algorithms for the analysis of randomly truncated data. • **FDA.USC**. Software for functional data analysis. • **geoR_NP**. Subroutines for estimating the variogram and non-parametric ordinary kriging.
MOSISOLID	• Resolution, by means of finite element methods, of contact models of elastic, viscoelastic and viscoplastic solids, including friction, adhesion, wear, heat transfer, plastification, materials with memory, etc.
varidis	• Mesh generation for the application of the finite element method.

3.6
Expertise in use of free and commercial software

In this section, free and commercial software that could be incorporated into mathematical technology on offer by the research groups is presented. The list below includes a brief description of each package, as well as the groups with proven expertise in its use.

Table 3.7 Free and commercial software (with a brief description) and research groups in the Consulting Platform with proven expertise in its use

ACTRAN GMFN, mat+i	Simulation in acoustics and vibro-acoustics.
AIMSUM PROMALS	Traffic simulator.
ALBERTA DDA	Free adaptive finite element software.
AMPL GIO, GIOPTIM, GNOM, GSO, MCS-UAB, modestya	Algebraic modelling tool for optimization problems.
ANSYS GMFN, GSC	Finite element package for general use in multiphysics.
ARCVIEW MODESMAN, TD-ULPGC	Package for geographical information and spatial analysis.
AUTOCAD EDANYA, varidis	CAD package. Architectural design.
AVS DDA	Graphic visualisation package.
BUGS (WINBUGS) MODESMAN	Bayesian statistical package.
CATIA mat+i, MOSISOLID	Set of computer applications for computer-aided design, manufacturing and engineering.

continued...

CENTAUR ADF	Non-structured mesh generator.
CGAL LOGRO	Computational geometry algorithm library.
Claroline EOPT, INFERES	Learning management system for e-learning and e-working.
CoCoA ACEIA	Symbolic computation language.
CoDaPack CODA	Statistical package for the analysis of compositional data.
CODE::BLOCKS MCS-UAB	Development environment for programs in C++.
COIN-OR EOPT, GSO	Solving linear and/or whole optimization problems.
Common Lisp Psycotrip	Programming language.
COMSOL Multiphysics **(before FEMLAB)** CAG, GMFN, GSC, KINETIC, M2NICA, MAI, mat+i, MCS-UAB, MOSISOLID, TTM	Modelling and analysis for virtual prototype of physical phenomena.
CPLEX EOPT, GIOPTIM, GNOM, GOMA, GOR, GPB97, GSO, InterTech, LOGRO, PROMALS, RUTYLO, RU-TYMETA	Integer and linear programming software.
C-XSC GIO	Interval analysis library.
DAMFLOW M2S2M	Hydrodynamic flow modelling.
Darttrainer Pro DEPREN	Movement analysis package.
DCPIP, **RES(2D3D)INV,** **EARTH-IMAGER** GMFN	Inversion programs in electrical tomography.

continued…

DERIVE	Symbolic computation language.
ACEIA	

DEV C++	Development environment for programs in C++.
ADF, DEPREN, GOR, KINETIC, MCS-UAB	

DIVINE	Seismic tomography program.
GMFN	

DLR TAU	Non-structured CFD 3D code.
ADF	

DSTool	Package for the analysis of dynamic systems.
DEPREN	

ECOLAB	Simulation in chemical kinetics.
mat+i	

ENSIGHT	Visualisation package.
mat+i	

EVIEWS	Econometry and time series analysis software.
MCS-UAB, MODES, modestya	

FIDAP	CFD: advanced material simulation.
MAI	

FLUENT	CFD.
ADF, MAI, mat+i	

FLUX	Simulation in electromagnetism.
mat+i	

FREEFEM	General purpose finite element package.
EDANYA, GMFN, M2S2M	

Freehand	Design program.
LOGRO	

GAMBIT	Preprocessor: geometry definition and mesh generation.
ADF, mat+i	

GAMS	Mathematical programming and optimization software.
EOPT, GPB97, GSO, LOGRO, MODESI, TD-ULPGC	

continued...

| **GAUSS** | Numerical problem solving in Statistics and Econometrics. |
| OEDgroup | |

| **GEOMETRY EXPRES-SIONS** | Dynamic geometry system. |
| ACEIA | |

| **GoogleMap** | Geographical information system. |
| DEPREN, GOMA | |

| **GRAV2D, GRAV3D, GRAVCAD** | Gravimetry inversion programs. |
| GMFN | |

| **GSLIB** | Geostatistical software library. |
| CODA, GMFN | |

| **I-DEAS** | Solid mechanics simulation by finite elements. |
| mat+i, MOSISOLID | |

| **KISTLER BIOWARE** | Package for analysis of force platform biomechanical data. |
| DEPREN | |

| **KNITRO** | Algebraic modelling tool for optimization problems. |
| modestya | |

| **LAPACK** | Linear algebra package. |
| AALN | |

| **LINDO (LINGO)** | Linear, nonlinear and integer programming software. |
| CYOPT, DEPREN, EOPT, GIO, GIOPTIM, GOMA, GOR, GPB97, GSO, MCS-UAB, OEDgroup, RUTYLO, RUTYMETA, TD-ULPGC, TTM | |

| **MAGMA** | Computer algebra system. |
| CG, MCS-UAB | |

| **MAPLE** | Symbolic computation package. |
| ACEIA, CAG, CG, CYOPT, DEPREN, EDANYA, EDnL, GAUCA, GPB97, INFERES, KINETIC, M2NICA, MCS-UAB, MODES, modestya, modsol, SSD, TAPO, TOREFA, varidis | |

| **MARC** | Nonlinear solid mechanics simulation by finite elements. |
| mat+i, MOSISOLID | |

| **MATCH VISION** | Game analysis package. |
| DEPREN | |

continued...

MATHEMATICA   Specialised package in numerical analysis and symbolic computation.

ACEIA, CAG, CG, DDA, DEPREN, EDANYA, EDnL, GAUCA, GIO, GMFN, HYPCHAOP, INFERES, InterTech, KINETIC, LOGRO, M2NICA, MCS-UAB, MODES, MODESI, modestya, modsol, MOSISOLID, OEDgroup, SSD, TAMI, TAPO, TOREFA,TTM

MATLAB   Numerical computation package with specific purpose toolboxes.

AALN, ADF, CAG, CG, CODA, CYOPT, DATFUN, DDA, DEPREN,EDANYA, GEUVA, GIOPTIM, GMFN, GNOM, GSC, GSO, HYPCHAOP, INFERES, InterTech, KINETIC, LOGRO, M2NICA, MAI, mat+i, MCS-UAB, MODES, MODESI, modestya, MOSISOLID, OEDgroup, TAMI, TAPO, TD-ULPGC, TOREFA, varidis

MAXIMA   Symbolic computation language.

ACEIA, modestya, TAPO

MEGAWAVE2   Image processing software.

TAMI

MIKE   Simulation in river and lake hydrodynamics.

GSC, mat+i

MINITAB   Statistical package.

CODA, MCS-UAB, MODES

MODFLOW, FEEFLOW   Hydrogeological modelling programs.

GMFN

MOODLE   Learning management system for e-learning and e-working.

GEUVA, GSO, INFERES

NAG   Numerical computation software.

CYOPT

NASTRAN   Finite element simulation in solid mechanics.

mat+i, MOSISOLID

NETGEN   2D and 3D Mesh Generator for finite elements.

M2NICA

Octave   MATLAB free software.

SSD, TTM

PATRAN   Pre- and post-processor for CAE simulation.

mat+i, MOSISOLID

PHOENIX   Finite volumes in fluid dynamics.

GMFN

continued…

PINACLE DEPREN	Video analysis package.
Polar DEPREN	Package for the analysis of physiological signals.
PRO PLANING PROFE-SIONAL DEPREN	Game analysis package.
PYTHON ADF, EDANYA, M2NICA, mat+i, MCS-UAB	Multi-paradigm programming language; providing various programming styles: programming aimed at objects, structured programming, functional programming and programming aimed at aspects.
R CG, CODA, DATFUN, EE, EOPT, GEUVA, GIOPTIM, GOMA, GSO, INFERES, MCS-UAB, MODES, MODESI, MODESMAN, modestya, OEDgroup, TAPO, TTM	Free statistical software.
S (S-PLUS) CODA, DATFUN, GEUVA, GOMA, MODES, MODESI, MODESMAN, modestya	Statistical package.
SAGE CG, M2S2M	Software for algebra and geometry experimentation.
SALOME M2NICA	Free CAD modelling software, finite element mesh generator and post-processing (with hdf5).
SAS CODA, EE, GEUVA, GOMA, GPB97, MCS-UAB	Statistical package.
SCILAB CYOPT, EDANYA, GNOM, TAMI	High-level programming language for scientific calculation, interactive and freely available in multiple operating systems.
SIMAIL MOSISOLID	Finite element mesh generator package.
SIMNET II MCS-UAB	Discrete simulation language.
Singular CG, SSD	Algebraic manipulator.

continued...

SPORT DRAW   Game analysis package.

DEPREN

SPSS   Data processing.

CODA, GEUVA, GOMA, GOR, GPB97, INFERES, MCS-UAB, MODES, MODESI, modestya, OEDgroup, TTM

STATA   Statistical package.

CODA, GOMA

STATGRAPHICS   Descriptive statistical package and quality control.

GEUVA, GOMA, GPB97, MODES, OEDgroup

STATISTICA   Statistical package.

GEUVA, GOMA, INFERES, modestya

SYSTAT   Statistical package.

DATFUN, modestya

TableCurve 2D and 3D   Statistical package, regression in 2 or 3 dimensions.

modestya

TACTIC PRO   Game analysis package.

DEPREN

TECPLOT   Graphic visualisation package.

ADF

The Geometer's Sketchpad   Dynamic geometry system.

ACEIA

Visual Studio.NET   Development environment.

GIOPTIM

VTK   3D visualization library.

EDANYA, M2NICA, mat+i

WEKA   Data mining software.

GIOPTIM, MODESI

WinQSB   Operational research models package.

GOR

continued…

wxWidgets	Library of components for the creation of GUIs.
MCS-UAB	
XPRESS	Mathematical programming.
GNOM, GOMA, GOR, GPB97, GSO, INFERES, modestya	
XPRESS-MP	Integer and linear problem optimizer.
GIO, GOR, GSO, INFERES	
YahooMap	Geographical information system.
GOMA	

3.7
Conclusions

The analysis and update of the information extracted from the survey included in this chapter have been a priority among all knowledge and technology transfer activities developed in the i-MATH project between 2007 and 2012. This study, along with the TransMATH Demand Map, has provided the basis of necessary information for the successful execution of almost all knowledge transfer activity in industrial mathematics carried out in Spain during those years.

The study has enabled us to:

- identify the most representative Spanish groups in industrial mathematics and create their portfolios of capacities and technologies;
- bridge the gap between Spanish mathematicians and companies, highlighting to industry potential areas of cooperation in which Spanish groups are particularly strong;
- unite the Spanish mathematical community working in knowledge transfer, thus improving its relationship with companies, through the promotion and identification of new possible collaborations and research lines which may be of interest to industry.

The map has also established the basis for the creation of two new structures aimed at fostering industrial mathematics in Spain:

- the Spanish Network for Mathematics and Industry, math-in (www.math-in.net), which aims to increase the presence of mathematical methods and techniques in the manufacturing sector by bringing together all Spanish research groups, which have an interest in mathematical knowledge transfer, with industry;
- the Technological Institute for Industrial Mathematics, ITMATI (www.itmati.com), whose primary objective is to enhance competitiveness and innovation in companies, operating as a centre of excellence in industrial mathematics on an international stage.

In this section, we summarise the main conclusions about this study related to the supply of mathematical techniques and expertise by all the 62 Spanish research groups in the Consulting Platform, which represents over 620 researchers and support staff working together for the benefit of industrial organisations.

3.7.1
Expertise in mathematical techniques

The joint expertise of the groups particularly stands out in the case of three specific areas of research, using the MSC classification. Firstly, the 90-xx MSC area, *Operations research and mathematical programming*, with 23 groups (37% of research groups in the Consulting Platform), to which are of particular importantance mathematical programming and optimization techniques. Secondly, the 62-xx MSC area, *Statistics*, with 19 groups (31%) with demonstrable relevant experience. Most noteworthy is the use of techniques such as data mining, time series or multivariate analysis. The third most important MSC research area is *Numerical analysis*, MSC area 65-xx, with 17 groups (27%), the most relevant areas of research being mathematical modelling and numerical simulation.

Techniques which stand out amongst those most widely used by groups in the Consulting Platform are techniques in fluid mechanics within CAD/CAE research (such as hydrodynamic flow simulation, computational fluid dynamics (CFD) or aerodynamic simulation); routes planning and optimization, and strategy and planning in logistics and transport, in the statistics and operations research area (ST/OR); and, regarding other mathematical techniques (OMT), development in software and computational applications.

Although the distribution of techniques encompasses the three main areas of expertise (CAD/CAE, ST/OR and OMT), the level of expertise in the field of statistics and operations research should be particularly emphasised due to the large number of groups participating in the survey showing an in-depth knowledge in this area of research.

3.7.2
Research activity aimed at knowledge transfer (2000–2010)

From the information taken from the survey for the period 2000 to 2010, it is clear that there was significant activity in knowledge transfer carried out by research groups. In total, more than 960 different types of collaborations with industrial partners have been identified, among them, the development of research projects in cooperation with companies, the execution of contracts for knowledge-intensive services to industry and the running of industrial training courses. It should be noted that the sample of groups that took part in the survey included the most active groups in the Spanish context of industrial mathematics, since they have all been active in mathematical transfer activities promoted by the i-MATH project. Moreover, data analysis shows that most of the activity is concentrated in a few groups with long experience in industrial mathematics.

If we make a breakdown by type of collaboration, during the period 2000–2010, the 62 research groups developed a total of 476 competitive research projects, which obtained funding through public calls at regional, national or international levels. Nevertheless, only 103 of them were supported, at least partially, by companies and industry bodies. Thus, the percentage of research projects in collaboration with companies versus the total research projects developed from 2000 to 2010 (only 22%) reflects the distance between research and industry, most research activity carried out by the groups during this period being conducted far away from the direct demands of companies.

However, the high number of direct contracts with companies and industry bodies held in that period is of significance. Groups have executed a total of 375 contracts for knowledge-intensive technology services, as well as a number of training courses for technology organisations or firms (121 courses). In this sense, it is important to highlight that some contracts have continued year on year, giving rise to long and fruitful relationships between the company and the research group: the former increases its profit applying innovative mathematical tools, whereas the latter transfers knowledge to industry and society and discovers new mathematical research lines with high (and often unknown) demand in business sectors.

Regarding the economic sectors in which contracts and training courses were most prevalent, *Public Administration*, *Energy & Environment* and *Metal & Machinery* are the most remarkable sectors. Furthermore, as far as the territorial distribution of the groups which developed direct contracts and training courses with industry is concerned, these knowledge transfer activities are concentrated in regions such as Galicia, Castile and León, Navarra, Valencia and Catalonia.

3.7.3
Consulting and training: expertise by sector

As for experience in each industrial sector of economic activity, the largest numbers of research groups with proven experience were found in *ICT*, *Energy & Environment* and *Metal & Machinery*. When performing an analysis by number of collaborative activities in each sector (considering contracts and training as a whole), *Public Administration*, *Energy & Environment*, and *Metal & Machinery* sectors stand out.

Considering both contracts and training courses, we observe that the *Public Administration* sector ranks first in both cases. The relevant representation of this sector is mainly due to the demand, from regional or national institutions, of services such as design and analysis of surveys, statistical data analysis or social studies.

The second largest sector where there is notable interaction in terms of contracts is the *Energy & Environment* sector; while in terms of training courses it is *Biomedicine & Health*. These results are explained by taking into account the demand for knowledge-intensive mathematical services by companies that operate primarily in the *Energy & Environment* sector, especially in mathematical modelling and numerical simulation, and statistical techniques aimed at controlling processes and making predictions. As for training courses, the *Biomedicine & Health* sector stands out, with a high demand for courses related to statistical data analysis, applied primarily by hospitals and health policy makers.

From the analysis of information by sector, it can also be concluded that during the period considered, groups focused a significant part of their research in matters of social and economic investigation in accordance with priorities established in national government research programs and corresponding regional governments. Such programs are often the main source of funding for groups' activities.

Collaborations between research groups and industry are, however, becoming ever more common, as the increment in the number of direct contracts and training courses from 2000 to 2008 shows (see Fig. 3.10 in Sect. 3.4.1.3). Unfortunately, the economic crisis has held this evolution back during recent years and in the final years of the study. Nevertheless, due to the willingness and disposition of firms and research groups to continue collaborating, it is expected that as the economic situation of companies improves, the current status will revert back to the positive trend observed pre-2008.

3.7.4
Software development and use

The development and use of software for the application of mathematical and statistical techniques is essential. Therefore, research groups developed a total of 111 software packages; of these, 35% have been transferred to companies or industrial organisations. Most of the software not being transferred was developed in the framework of individual research projects carried out by the groups. In terms of expertise in handling free or commercial software, the groups show experience in using more than 95 software packages, among them the most widespread mathematical and statistical software.

At this point, it is important to note the increasing demand for training courses related to the use of mathematical and statistical packages, both free and commercial software and customised modules were developed by the research groups, most of them adapted to the specific needs and requirements of the company or client.

3.7.5
Expertise vs. supply

When analysed, for each sector of activity, the number of groups with proven experience in mathematical technology transfer (which have developed projects, contracts, training courses or other knowledge transfer activities) versus the number of groups that offer services, the available supply of all groups significantly exceeds the proven experience. This reflects, on the one hand, the variety of fields of possible application of the techniques developed, which can be used across different sectors and, on the other, the gap between the groups' capabilities and resources and that proportion of them that is finally concreted in collaborations with industry.

Although this difference is observed in all sectors without exception, it should be noted, for example, the very high percentage of groups that offer knowledge transfer in *Construction* without having proven experience in this sector (44%): 16 groups offer services, but only 7 have expertise. On the contrary, 92% of groups which offer technological services in the *ICT* sector have developed some kind of knowledge transfer activity with industry in this field.

3.7.6
Demand vs. supply

In Chap. 2 a global overview of the demand for mathematical technology in Spanish industry was presented, while Chap. 3 has been devoted to the supply of mathematical technology by Spanish research groups. Below, Fig. 3.11 compares the distribution by sector of the 286 firms which stated their interest in mathematical services or qualified mathematicians or statisticians (see Fig. 2.20, Sect. 2.2.3.3) and the distribution by sector of the technological supply offered by the research groups in the Consulting Platform (see Table 3.6, Sect. 3.4.2). Bear in mind that the *Public Administration* sector had been discarded in the demand survey, while *Technical Services* was not considered for technological supply.

Analysis by sector shows that the highest percentages of companies interested in mathematical services correspond to *ICT*, *Metal & Machinery* and *Technical Services* (14% each), whereas the majority of mathematical supply is focused in the *ICT* (19%), *Energy & Environment* (12%) and *Public Administration* (12%) sectors.

In general, it seems that technological supply and demand are reasonably balanced across economic sectors, although the demand for technological services is higher than supply in some of them, for example, *Construction*, *Economics & Finance*, *Food*, *Management & Tourism*, and *Metal & Machinery*. Correspondingly, *Biomedicine & Health*, *Energy & Environment*, *ICT*, and *Logistics & Transport* stand out among the economic sectors where the percentage of technological services offered exceeds the percentage of firms with mathematical needs.

Special attention must be paid to the *Technical Services* sector, since this field represents the 14% of firms which demand mathematical technology but, for the time being, where research groups do not supply specific mathematical services. This fact suggest that efforts must be made to put firms in the *Technical Services* sector in contact with research groups in order to transfer mathematical techniques for application in this field.

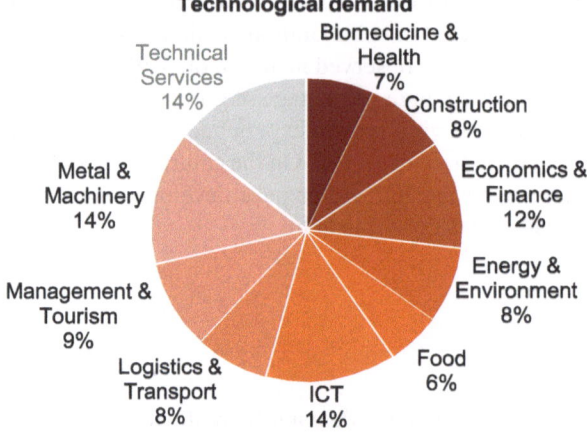

Fig. 3.11 Distribution of technological demand and technological supply by sector. Note: the *Technical Services* sector was only considered in the TransMATH Demand Map, while the *Public Administration* sector was only taken into account for the TransMATH Supply Map

References

Lery T, Primicerio M, Esteban MJ, Fontes M, Maday Y, Mehrmann V, Quadros G, Schilders W, Schuppert A, Tewkesbury H (eds) (2011) European success stories in industrial mathematics. Springer, Heidelberg

Quintela P, Sánchez MT, Martínez A, Parente G (2011) TransMATH: Investigadores en matemáticas para dar soluciones innovadoras. Project Ingenio Mathematica (i-MATH). http://www.i-math.org/files/File/documentos/mapa_consulting.pdf. Accessed 13 April 2011

Quintela P, Parente G, Sánchez MT, Fernández AB (2012) Soluciones Matemáticas para Empresas Innovadoras. Catálogo de Servicios Ofertados por Investigadores Españoles. McGraw-Hill, Madrid

Index by research group

Index by business sector

B

Biomedicine & Health, 5
    Demand, 8–11, 13–15, 17, 19, 26, 29, 31–40
    Supply, 89, 91, 96–98, 106, 109, 116, 127, 128, 132, 134, 137, 153–155

C

Construction, 5
    Demand, 8–11, 13–15, 17–19, 25, 26, 29, 31–34, 36, 37, 39, 40
    Supply, 91, 96–98, 102, 109, 116, 125, 128, 138, 155

E

Economics & Finance, 5
    Demand, 8–11, 13–17, 19, 25, 26, 28, 29, 31–34, 36, 37, 39, 40, 42, 43
    Supply, 91, 96, 97, 98, 101, 104, 122, 127, 134, 138, 155
Energy & Environment, 5
    Demand, 8–11, 13–19, 26, 29, 31–34, 36, 37, 39–41, 43
    Supply, 89, 91, 93, 96–98, 101, 106, 120, 122, 127, 128, 139, 153, 155

F

Food, 5
    Demand, 8–11, 13–15, 17, 19, 26, 29–34, 36, 37, 39–41
    Supply, 89, 91, 96–98, 101, 112, 120, 127, 128, 140, 155

I

ICT, 5
    Demand, 8–11, 13–17, 19, 26, 28, 29, 31–34, 36, 37, 39–42

    Supply, 91, 96–98, 100, 101, 103, 105, 114, 115, 125, 127, 128, 130, 134, 140, 153, 155

L

Logistics & Transport, 5
    Demand, 8–11, 13–15, 17, 19, 26, 29, 31–37, 39–41
    Supply, 81, 89, 91, 96–98, 103, 105, 106, 109, 110, 113, 117, 118, 127, 128, 130, 133, 141, 155

M

Management & Tourism, 5
    Demands 8–11, 13–15, 17, 19, 25, 26, 29–34, 36, 37, 39, 40
    Supply, 91, 96–98, 100, 101, 105, 116, 121, 131, 133, 134, 141, 155
Metal & Machinery, 5
    Demand, 8–11, 13–19, 25, 26, 29, 31–34, 36–43
    Supply, 89, 91, 93, 96–98, 103, 110, 111, 112, 116, 117, 119, 122, 123, 141, 153, 155

P

Public Administration, 5
    Demand, 8
    Supply, 91, 93, 96–98, 100, 106, 109, 114, 120, 122, 123, 127, 128, 143, 153, 155, 156

T

Technical Services, 5
    Demand, 8–11, 13–19, 25, 26, 28, 29, 31–34, 36–42
    Supply, 155, 156

1. L. Cinquini, A. Di Minin, R. Varaldo (Eds.)
   Nuovi modelli di business e creazione di valore: la Scienza dei Servizi
   2011, xvi+254 pp, ISBN 978-88-470-1844-0

2. H. Chesbrough
   Open Services Innovation – Competere in una nuova era
   2011, xiv+216 pp, ISBN 978-88-470-1979-9

3. G. Conti, M. Granieri, A. Piccaluga
   La gestione del trasferimento tecnologico. Strategie, modelli e strumenti
   2011, x+218 pp, ISBN 978-88-470-1901-0

4. M. Bianchi, A. Piccaluga (Eds.)
   La sfida del trasferimento tecnologico: le Università italiane
   si raccontano
   2012, xviii+194 pp, ISBN 978-88-470-1976-8

5. M. Granieri, A. Renda
   Innovation Law and Policy in the European Union. Towards
   Horizon 2020
   2012, xii+198 pp, ISBN 978-88-470-1916-4

6. P. Quintela, A.B. Fernández, A. Martínez, G. Parente, M.T. Sánchez,
   TransMath. Innovative Solutions from Mathematical Technology
   2012, xii+162 pp, ISBN 978-88-470-2405-2

http://www.springer.com/series/10062

Editor at Springer:
F. Bonadei
francesca.bonadei@springer.com